*AuthorHouse™ UK Ltd.*
*500 Avebury Boulevard*
*Central Milton Keynes, MK9 2BE*
*www.authorhouse.co.uk*
*Phone: 08001974150*

*First published by AuthorHouse     12/23/2010.*

*ISBN: 978-1-4520-9556-1 (sc)*
*ISBN: 978-1-4520-9551-6 (e)*

*This book is printed on acid-free paper.*

# GRADUATED ALGEBRA

## FOR THE YOUNG SCHOLAR

# A.J Aseervatham

# 2011

## PREFACE

A thorough understanding of School Algebra is essential for a student to do well in all other areas of mathematics. This book, "Graduated Algebra" aims to achieve this by providing graded exercises in all relevant basic algebra topics to help hone the knowledge of the young scholars.

This book is designed to assist students who have had difficulty with the fundamental concepts of algebra and are struggling with it in higher grades.

By systematically attempting each exercise in this book, the students will begin to enjoy Algebra. Answers are provided at the end of each exercise to facilitate easy checking.

I was prompted to write this book after seeing a lot of students struggling to get the right answers to mathematical questions because of their poor knowledge of Algebra. I am confident that the contents of this book will help young people to excel in mathematics.

I thank Antonio Mahadevan for kindly agreeing to format my work and for double checking the answers for accuracy.


Al Aseervatham                                    January 2011


## DEDICATION

I dedicate this book to all those students around the world to whom I had the pleasure of teaching mathematics. The countries where I taught include Sri Lanka, United Kingdom, Ghana, Nigeria, Zambia, Malawi and Australia.

# TABLE OF CONTENTS

| | | |
|---|---|---|
| 1. | INTRODUCTION TO ALGEBRA | 1 |
| 2. | ADDITION | 7 |
| 3. | SUBTRACTION | 10 |
| 4. | ADDITION AND SUBTRACTION | 12 |
| 5. | SIMPLE MULTIPLICATION | 14 |
| 6. | ADVANCED MULTIPLICATION | 16 |
| 7. | SIMPLE DIVISION | 25 |
| 8. | LONG DIVISION | 27 |
| 9. | MULTIPLICATION AND DIVISION | 31 |
| 10. | THE FOUR RULES | 34 |
| 11. | REMOVAL OF BRACKETS | 37 |
| 12. | SQUARE ROOTS AND CUBE ROOTS | 39 |
| 13. | SURDS AND INDICES | 43 |
| 14. | LOGARITHMS | 51 |
| 15. | SUBSTITUTION | 56 |
| 16. | FACTORISATION | 59 |
| 17. | HIGHEST COMMON FACTOR | 70 |
| 18. | LOWEST COMMON MULTIPLE | 73 |

| | | |
|---|---|---|
| 19. | FRACTIONS | 76 |
| 20. | SIMPLE EQUATIONS | 84 |
| 21. | SIMPLE EQUATIONS WITH FRACTIONS | 88 |
| 22. | SURDS, MODULUS, EXPONENTS AND LOGS | 94 |
| 23. | SIMPLE INEQUATIONS | 103 |
| 24. | CHANGING THE SUBJECT OF A FORMULA | 107 |
| 25. | SIMULTANEOUS EQUATIONS | 111 |
| 26. | QUADRATIC EQUATIONS | 119 |
| 27. | POLYNOMIAL FUNCTIONS | 126 |
| 28. | PARTIAL FRACTIONS | 137 |
| 29. | PATTERNS, SEQUENCES AND SERIES | 144 |
| INDEX | | 169 |

# CHAPTER 1

## INTRODUCTION TO ALGEBRA

When we do problems in arithmetic we are dealing with numbers using the four basic operations: addition, subtraction, multiplication and division. Instead of numbers we can use any letter in the English alphabet to represent a number.  For example, we can let the letter 'x' represent any number.

| |
|---|
| Since x can represent any number we call 'x' a **variable** |

We can also use any letter of the alphabet as the variable. We can then define a variable as a symbol that represents any value.

| |
|---|
| If a variable represents a specific value then it is no more a variable.  We call it a **'constant'.** |

Suppose we have three variables x, y and z and a constant number 3.

We can add the constant to the variables, subtract the constant from the variables and multiply or divide the variables by the constant.

- **Addition** gives the results: x +3, y + 3, z + 3

- **Subtraction** gives the results: x – 3, y – 3, z – 3

- **Multiplication** gives the results: 3x, 3y, 3z or 3.x, 3.y, 3.z

- **Division** gives the results: $\dfrac{x}{3}, \dfrac{y}{3}, \dfrac{z}{3}$ or $\dfrac{1}{3}x, \dfrac{1}{3}y, \dfrac{1}{3}z$

Multiplication is expressed as 3 × x, but it is customary to write it as 3x. This means that the constant is multiplying the variable.

We write the number first and the variable next to it (we don't write x3, y3, z3). It is also acceptable to write these expressions as 3.x, 3.y and 3.z.

## THE COEFFICIENT OF A VARIABLE

When a constant multiplies a variable we write the constant in front of the variable. This constant number is then referred to as the **coefficient** of the variable.

**Example:**

In 5x, 4y and 9z;

- 5 is the coefficient of x
- 4 is the coefficient of y
- 9 is the coefficient of z

## MULTIPLYING A VARIABLE BY ITSELF

The result of multiplying the variable x by x is written as $x^2$.

That is; $x \times x = x^2$  (read x to the power 2)

Similarly,

- $x \times x \times x = x^3$  (read x to the power 3)
- $x \times x \times x \times x = x^4$  (read x to the power 4)
- $y \times y \times y \times y \times y = y^5$  (read y to the power 5)

## MULTIPLYING VARIABLES AND CONSTANTS

When two variables x and y are multiplied together the result is written as xy. When three variables x, y and z are multiplied together the result is written as xyz, and so on.

Terms such as xy and xyz may be called **product variables**.

**Examples:**

The expression $x^2y$ means that x is multiplied by x and then by y.

That is, $x^2y = x \times x \times y$   (variable x is repeated)

$4xy^2 = 4 \times x \times y \times y$   (4 is a constant, variable y is repeated)

$7y^3z^2 = 7 \times y \times y \times y \times z \times z$   (7 is a constant, variables y and z are repeated)

Product variables such as $4xy^2$ and $7y^3z^2$ are called **complex variables.**

The result of multiplying two or more product or complex variables is expressed as a single complex variable as shown in the following examples.

**More Examples:**

- $xy \times x^2y = x \times y \times x \times x \times y = x^3y^2$

- $2xy^2 \times 3x^3y^2 = 2 \times x \times y \times y \times 3 \times x \times x \times x \times y \times y = 6x^4y^4$

## ALGEBRAIC EXPRESSIONS

An algebraic expression is a combination of variables involving one or more of the arithmetical operations of addition, subtraction, multiplication and division.

Any part of an expression is referred to as a **term.**

**Examples:**

$x + y, \ x^2 + y^2, \ -x^2 - 2x, \ 3x^2 + 4x - 8, \ x^2y + 3yz, \ \dfrac{2x - 4y}{3x + 5y}$

In the expression:

- $x + y$, x is a term and y is a term

- $x^2 + y^2$, $x^2$ is a term and $y^2$ is a term

- $-x^2 - 2x$, $-x^2$ is a term and $-2x$ is a term

- $3x^2 + 4x - 8$, $3x^2$ is a term $4x$ is a term and $-8$ is a term

- $x^2y + 3yz$, $x^2y$ is a term and $3yz$ is a term

- $\dfrac{2x - 4y}{3x + 5y}$, $2x - 4y$ is a term and $3x + 5y$ is a term

If the first term of an expression is written without a negative sign it is assumed to have a positive sign in front of it. This is the appropriate method of writing a positive first term in an expression.

3

# LIKE AND UNLIKE TERMS

Terms involving same variables and same powers are **like terms**.

**Examples:**

- $5x$, $-2x$, $8x$   - $-3x^2$, $4x^2$, $-7x^2$   - $9x^2y^2$, $-6x^2y^2$, $-x^2y^2$

**Unlike terms** are terms other than like terms.

**Examples:**

- $6x$, $3y$, $-5z$   - $4yz$, $3xy$, $10x^2$   - $2x^3y$, $5x^2y^2$, $7xy$

# CONSTRUCTING ALGEBRAIC EXPRESSIONS

Since we can represent unknowns by variables, we can construct expressions for a value that we want to calculate.

For example if we let the length and breadth of a rectangle to be x cm and y cm respectively, we can construct algebraic expressions for its perimeter and area as follows:

Perimeter of the rectangle = x + x + y + y cm.

Area of the rectangle = Length × Breadth = x × y cm$^2$.

**EXERCISE 1.1**----------------------------------------------------------------------------

Write the expanded version of each of the following:

| | | | |
|---|---|---|---|
| 1.  $2x$ | 3.  $3x^3$ | 5.  $x^3y^4$ | 7.  $-4a^2b^3c^2$ |
| 2.  $x^2$ | 4.  $x^2y$ | 6.  $5m^4n^2$ | 8.  $8p^3q^2$ |

--------------------------------------------------------------------------------------------------

| Solutions: Exercise 1.1 | |
|---|---|
| **1.** $2 \times x$ | **5.** $x \times x \times x \times x \times y \times y \times y \times y$ |
| **2.** $x \times x$ | **6.** $5 \times m \times m \times m \times m \times n \times n$ |
| **3.** $3 \times x \times x \times x$ | **7.** $-4 \times a \times a \times b \times b \times b \times c \times c$ |
| **4.** $x \times x \times y$ | **8.** $8 \times p \times p \times p \times q \times q$ |

**EXERCISE 1.2**----------------------------------------------------------------------------

Write the simplified version of each of the following:

1. $x \times y \times 2$

2. $3 \times a \times 2 \times b$

3. $4 \times x \times -3 \times x$

4. $x^2 \times 3 \times x \times y^2 \times 2$

5. $-5 \times c^2 \times b^3 \times 2c$

6. $7 \times m \times 2 \times -n \times n$

7. $p \times 2q \times 3r$

8. $x^2 y^3 \times 3xy^2 \times xy$

---

| Solutions: Exercise 1.2 | | | |
|---|---|---|---|
| **1.** $2xy$ | **3.** $-12x^2$ | **5.** $-10b^3c^3$ | **7.** $6pqr$ |
| **2.** $6ab$ | **4.** $6x^3y^2$ | **6.** $-14mn^2$ | **8.** $3x^4y^6$ |

**EXERCISE 1.3**----------------------------------------------------------------------------

For each of the following algebraic expressions state each term involving a variable and its coefficient. What is the constant term if any?

1. $2a + 3b$

2. $3x - 4y + 3$

3. $5x^2 + 3x - 6$

4. $y^2 + 2y - 9$

5. $m^2 - m - 6$

6. $3x^2 - 5y^2$

7. $a^2 - 8b^2$

8. $x^3 + 4x^2 - x + 5$

---

| Solutions: Exercise 1.3 | |
|---|---|
| **1.** $2a$ and $3b$ (2, 3) | **5.** $m^2$ and $-m$ (1, $-1$), C = $-6$ |
| **2.** $3x$ and $-4y$ (3 and $-4$), C = 3 | **6.** $3x^2$ and $-5y^2$ (3, $-5$) |
| **3.** $5x^2$ and $3x$ (5, 3), C = $-6$ | **7.** $a^2$ and $-8b^2$ (1, $-8$) |
| **4.** $y^2$ and $2y$ (1, 2), C = $-9$ | **8.** $x^3$, $4x^2$ and $-x$ (1, 4, $-1$), C = 5 |

**EXERCISE 1.4**----------------------------------------------------------------------------

In each of the following, pick the odd one out:

1.    $x, -4x, y$

2.    $2x^2, 4y, -6y$

3.    $3x, -x, 4y^2$

4.    $5x, 9x, -6x^2$

5.    $ab^2, a^2b, -3ab^2$

6.    $4a^3, 4x^3, 6a^3$

7.    $xy^3, x^3y, 2x^3y$

8.    $m^2n, 3n^2m, mn^2$

---

| Solutions: Exercise 1.4 | | | |
|---|---|---|---|
| **1.** $y$ | **3.** $4y^2$ | **5.** $a^2b$ | **7.** $xy^3$ |
| **2.** $2x^2$ | **4.** $-6x^2$ | **6.** $4x^3$ | **8.** $m^2n$ |

**EXERCISE 1.5**---------------------------------------------------------------------------------

Simplify the following algebraic expressions by adding the like terms together:

1. $x + 3y + 4x$

4. $a^2 + ab^2 + 3a^2 - 4ab^2$

7. $ax^2 + bx - c + 4ax^2 - 6c$

2. $3x - 4y - 2y + 2x$

5. $x^2 - 3xy + xy - 2x^2$

8. $3xy^2 + 4xy - 4x^2y - 2xy^2 + 8$

3. $2x^2 + y^2 - 2y^2 + x^2$

6. $2x - y^2 + 2y + 3y^2 + 5 - 2x$

---

| Solutions: Exercise 1.5 | | | |
|---|---|---|---|
| **1.** $5x + 3y$ | **3.** $3x^2 - y^2$ | **5.** $-x^2 - 2xy$ | **7.** $5ax^2 + bx - 7c$ |
| **2.** $5x - 6y$ | **4.** $4a^2 - 3ab^2$ | **6.** $2y^2 + 2y + 5$ | **8.** $-4x^2y + 4xy + xy^2 + 8$ |

**EXERCISE 1.6**---------------------------------------------------------------------------------

Construct the appropriate algebraic expressions for each of the following:

1. If a bag of oranges cost $x, how much will 4 bags of oranges cost?

2. If the three sides of a triangle are a cm, b cm and c cm respectively, what is the perimeter of the triangle?

3. If I divide $y among 3 people equally, how much will each person get?

4. If a square has a side of length x cm, what is its area?

5. I went to a coffee shop and had a sandwich for $x and a coffee for $y. How much did I spend?

6. A man bought 12 soup packets at $x each and 6 cans of mixed fruits at $y per can. If he paid with a $50 note, how much change would be given?

7. If I travelled x km at 60 km per hour and y km at 80 km per hour, how long did I travel for?

8. The wages paid for workers vary according to their skills. A skilled worker is paid $p per hour, a semi-skilled worker is paid $q per hour and an unskilled worker is paid $r per hour. On a particular day, there were 8 skilled workers, 6 semi skilled workers and 20 unskilled workers. What was the total wages paid for an 8 hour day?

---

| Solutions: Exercise 1.6 | | | |
|---|---|---|---|
| **1.** $4x | **3.** $\frac{y}{3}$ | **5.** $(x + y)$ | **7.** $(\frac{x}{60} + \frac{y}{80})$hours |
| **2.** $(a + b + c)$cm | **4.** $x^2$ cm$^2$ | **6.** $(50 - 12x - 6y)$ | **8.** $8(8p + 6q + 20r)$ |

# CHAPTER 2

When adding multiple algebraic expressions, like terms must be added together. This is facilitated by writing the like terms from each expression in the same column, one below the other as shown in the following examples.

**Example 1:**

Add the following algebraic expressions:

$2a + 3b - 4c$, $3a + 5c$, $4b + 3c$

Arrange the expressions one below the other with like terms in the same column, and add:

$$
\begin{array}{c|c|c}
2a & +3b & -4c \\
3a & & +5c \\
& 4b & +3c \\
\hline
5a & +7b & +4c
\end{array}
$$

**Example 2:**

Add the following algebraic expressions:

$5x^3 - 2x$, $3x^2 + 4x$, $x^4 - 6x^3$, $6x^4 - 2x^2$

Arrange the expressions one below the other with like terms in the same column, and add:

$$
\begin{array}{c|c|c|c}
& 5x^3 & & -2x \\
& & 3x^2 & +4x \\
x^4 & -6x^3 & & \\
6x^4 & & -2x^2 & \\
\hline
7x^4 & -x^3 & +x^2 & +2x
\end{array}
$$

**Example 3:**

Add the following algebraic expressions:

$2a^2b + a^3 + 3ab^2 + b^3$, $a^3 + 3b^3 + 2a^2b$, $2b^3 - 5ab^2 + 3a^3$

Arranging each of the given expressions in descending powers of 'a', ascending powers of 'b' and the like terms in the same column one below the other we have:

| $a^3$ | $+2a^2b$ | $+3ab^2$ | $+b^3$ |
|---|---|---|---|
| $a^3$ | $+2a^2b$ | | $+3b^3$ |
| $3a^3$ | | $-5ab^2$ | $+2b^3$ |
| $5a^3$ | $+4a^2b$ | $-2ab^2$ | $+6b^3$ |

**EXERCISE 2.1** --------------------------------------------------------------------------

Add the following algebraic expressions together:

1.  $3a + 4b - c$;  $a - b + 3c$;  $2a - 3b - c$

2.  $5x + 2y - 7z$;  $3x - 3y + 4z$;  $x - 3y + z$

3.  $3ab + 4bc - 5ac$;  $ab - 3bc + 2ac$;  $-ab + bc + 3ac$

4.  $4lm - 3mn + ln$;  $-3lm + 2mn - ln$;  $5lm - mn + 3ln$

5.  $x^2 + 2x + 3$;  $3x^2 - 5x - 5$;  $3x^2 - 6x + 7$

6.  $3a^2 + 3a + 4$;  $2a^2 - 5a + 7$;  $-a^2 + 4a - 9$

7.  $m^3 + 2m^2 + m$;  $-2m^3 + m^2 - 5m$;  $m^3 - m^2 + 2m$

8.  $p^3 + 2p^2q + 5$;  $-3p^3 + pq^2 - 3$;  $2p^3 - pq^2 + 2p^2q + 2$

--------------------------------------------------------------------------

| Solutions: Exercise 2.1 | | | |
|---|---|---|---|
| **1.** $6a + c$ | **3.** $3ab + 2bc$ | **5.** $7x^2 - 9x + 5$ | **7.** $2m^2 - 2m$ |
| **2.** $9x - 4y - 2z$ | **4.** $6lm - 2mn + 3ln$ | **6.** $4a^2 + 2a + 2$ | **8.** $4p^2q + 4$ |

Add the following algebraic expressions together:

1.     $a^4 + 8a^3$;   $-6a^3 + 2a^2$;   $-4a^2 - 3a$

2.     $x^3 + x - 2x^2 - 5$;   $3x^2 + 5x + 2$;   $x^3 + 3x^2 - 4x$

3.     $1 - 3y + y^2 - 3y^3$;   $2 - 2y^2 + y^3$;   $3y + y^2 - 4y^3$

4.     $m^3 + 2m^2$;   $5m^2 - 7m$;   $m^3 + m^2 + 1$

5.     $-2x^2y + 5y^3 - 1$;   $4 + 3xy^2 - 4x^2y$;   $-2 - 3x^2y - y^3$

6.     $2a^2b - 2a^3 + 2ab^2 - 2b^3$;   $b^3 - 3a^2b$;   $-5a^2b + 4ab^2$

7.     $a^3 - 3b^3 + 3ab^2$;   $4a^2b - 5ab^2 - 2a^3$;   $6a^2b - a^3 + ab^2$

8.     $3p^3 - 2p^2q + 2$;   $q^3 - 3pq^2 + 4p^2q - 5$;   $2p^3 - 3pq^2 + 4$

----------------------------------------------------------------------------------------------

| Solutions: Exercise 2.2 | |
|---|---|
| **1.** $a^4 + 2a^3 - 2a^2 - 3a$ | **5.** $1 - 9x^2y + 3xy^2 + 4y^3$ |
| **2.** $2x^3 + 4x^2 + 2x - 3$ | **6.** $-2a^3 - 6a^2b + 6ab^2 - b^3$ |
| **3.** $-6y^3 + 3$ | **7.** $-2a^3 + 10a^2b - ab^2 - 3b^3$ |
| **4.** $2m^3 + 8m^2 - 7m + 1$ | **8.** $5p^3 + 2p^2q - 6pq^2 + q^3 + 1$ |

# CHAPTER 3

## SUBTRACTION

When subtracting one algebraic expression from another, like terms can only be subtracted. This is facilitated by writing the like terms from each expression on the same column, one below the other and in some order as shown in the following examples.

- Subtract $13x - x^2$ from $17x - 4x^2$

Arrange the like terms of the expressions one below the other and **subtract**:

$$
\begin{array}{c|c}
17x & -4x^2 \\
13x & -x^2 \\
\hline
4x & -3x^2
\end{array}
$$

- Subtract $4x^2y + 2x^3y^2 - xy^2$ from $2xy^2 - 5x^2y + 3x^3y^2$

Arrange the expressions in the order of the descending powers of x and like terms one below the other, and **subtract**:

$$
\begin{array}{c|c|c}
3x^3y^2 & -5x^2y & +2xy^2 \\
2x^3y^2 & +4x^2y & -xy^2 \\
\hline
x^3y^2 & -9x^2y & +3xy^2
\end{array}
$$

## EXERCISE 3.1----------------------------------------------------------------------

Subtract:

1. $2x - x^2$ from $x^2 - 3x$

2. $2a^2 - b^2$ from $a^2 + b^2$

3. $4x^2 - 3$ from $3x^2 - 2$

4. $2 - 3p^3$ from $5p^3 - 1$

5. $x^4 - 2x^2$ from $3x^2 - x^4$

6. $a^2 + a$ from $3a^2 - 1$

7. $2l^2 - 3m^2$ from $3l^2 - 2m^2$

8. $x^3 - 2x^2$ from $7x^3 - 4x^2$

-----------------------------------------------------------------------------------------

| Solutions: Exercise 3.1 | | | |
|---|---|---|---|
| **1.** $2x^2 - 5x$ | **3.** $-x^2 + 1$ | **5.** $-2x^4 + 5x^2$ | **7.** $l^2 + m^2$ |
| **2.** $-a^2 + 2b^2$ | **4.** $8p^3 - 3$ | **6.** $2a^2 - a - 1$ | **8.** $6x^3 - 2x^2$ |

**EXERCISE 3.2**----------------------------------------------------------------

Subtract the second expression from the first expression in each of the following:

1. $3a + 2b - c$;   $a + 3b + c$

2. $3x + y - 2z$;   $x - 3y + 3z$

3. $-xy - 3xz + 2yz$;   $xy - 2yz$

4. $2ab + ac$;   $-ab + bc - ac$

5. $5xy - 3zx$;   $2xy - yz - 3zx$

6. $ab + 4bc - 3abc$;   $2abc - 3ab$

7. $2xy - 2yz + 2zx$;   $-xy + 2yz - 3zx$

8. $7pq - 2qr + 5rs$;   $3qr + 5pq - 2rs$

| Solutions: Exercise 3.2 | |
|---|---|
| **1.** $2a - b - 2c$ | **5.** $3xy + yz$ |
| **2.** $2x + 4y - 5z$ | **6.** $-5abc + 4ab + 4bc$ |
| **3.** $-2xy - 3xz + 4yz$ | **7.** $3xy - 4yz + 5zx$ |
| **4.** $3ab + 2ac - bc$ | **8.** $2pq - 5qr + 7rs$ |

**EXERCISE 3.3**----------------------------------------------------------------

Subtract the first expression from the second expression:

1.   $5a^3 + 3a^2b - 3ab^2 + 4b^3$;   $6a^3 - 3a^2b + 4ab^2 + 3b^3$

2.   $-x^3 + 2xy^2 - 3x^2y - y^3$;   $2x^3 + 4x^2y + 3xy^2 + 2y^3$

3.   $5m^3 - 4mn^2 + n^3$;   $2m^3 + 7n^3 - 3m^2n + 5mn^2$

4.   $3a^3 + 2a^2b + 4ab^2 - b^3 + 7$;   $5a^3 + 3a^2b - 4 + 7ab^2 - 2b^3$

5.   $-3p^3 - 4p^2q - 2pq^2 + 2q^3$;   $-9p^3 + 11p^2q - 7pq^2 + q^3$

6.   $4ab^4 + 3a^2b^3 - 2a^3b^2 - a^5$;   $a^2b^3 - 3ab^4 + 5a^3b^2$

7.   $4ax - 2a^2 - 3b^2 + 6$;   $7ax - 3a^2 + 9 - 11b^2$

8.   $5a^3b + 9ab^3 - a^4 - 6b^4 - 3a^2b^2$;   $8ab^3 + 2a^3b - 4b^4$

| Solutions: Exercise 3.3 | |
|---|---|
| **1.** $a^3 - 6a^2b + 7ab^2 - b^3$ | **5.** $-6p^3 + 15p^2q - 5pq^2 - q^3$ |
| **2.** $3x^3 + 7x^2y + xy^2 + 3y^3$ | **6.** $a^5 + 7a^3b^2 - 2a^2b^3 - 7ab^4$ |
| **3.** $-3m^3 - 3m^2n + 9mn^2 + 6n^3$ | **7.** $-a^2 + 3ax - 8b^2 + 3$ |
| **4.** $2a^3 + a^2b + 3ab^2 - b^3 - 11$ | **8.** $a^4 - 3a^3b + 3a^2b^2 - ab^3 + 2b^4$ |

# CHAPTER 4

## ADDITION AND SUBTRACTION

It is now appropriate to do some calculations involving both addition and subtraction. The expressions need to be arranged in an order that will simplify the calculations.

### EXERCISE 4.1

1. Subtract $3a^2 + 2a + 1$ from $2a^3$ and to the result add $4a^2 + 3a + 1$.

2. From the sum of $x^2 + 2x + 7$ and $x^3 - 2x^2 + 3x + 2$ subtract $x^4 - x - 1$.

3. Add the sum of $2x - x^2$ and $-5x^3 + 1$ to the result obtained by subtracting $4x^3$ from $x - 2x^2 + 1$.

4. Subtract the sum of $3m - 5 + 6m^2$ and $3m^2 - 3 - 2m^3$ from $4m^3 + 3m - 1$.

5. From the sum of $3p^2 - 4pq + 2q^2$ and $p^3 - 3q^2 + 1$, subtract the sum of $4p^3 - 2pq - p^2$ and $3 + 5q^2 - 5p^3$.

6. From $3a^3 + 2a^2 - 8$ subtract the sum of $4 + 3a^2 + a^3$, $a^3 - a^2 - 9$ and $a^3 - 4a^2 + 7$.

7. Subtract from $2m^3 - 3mn^2 + 4$, the result of subtracting $1 - m^3$ from zero.

8. Add to the sum of $4lm - l^2 - 3m^2$ and $2l^2 + 5m^2$, the result of subtracting $l^2 - m^2$ from $3lm - 4m^2$.

---

| Solutions: Exercise 4.1 | |
|---|---|
| **1.** $2a^3 + a^2 + a$ | **5.** $2p^3 + 4p^2 - 2pq - 6q^2 - 2$ |
| **2.** $-x^4 + x^3 - x^2 + 6x + 10$ | **6.** $4a^2 - 10$ |
| **3.** $-9x^3 - 3x^2 + 3x + 2$ | **7.** $m^3 - 3mn^2 + 5$ |
| **4.** $6m^3 - 9m^2 + 7$ | **8.** $7lm - m^2$ |

1.  There are $9x^2 - 3xy + 4y^2$ male students and $2y^2 - 5xy$ female students in a classroom. If $3y^2 - 4x^2 - 2xy$ students left the classroom, how many students would remain?

2.  In a meeting convened by the four religious groups, Buddhists, Christians, Hindus and Muslims, there were $3m^2 - 2m + 8$ Buddhists, $2m^2 - 2$ Hindus and $2m - 3 - m^2$ Muslims. If the total attendance of the meeting was $7m^2 + 3$, how many were Christians?

3.  In School A, there are $x^3 - 4x^2 + 5x + 6$ male students and $3x^2 - x + 4$ female students. In School B, there are $2x - 3$ less male students and $x^2 + 1$ more female students than in School A. How many students are in School B?

4.  A fruit seller had $4p^2 - q^2 + pq + 1$ bags of oranges. He bought an additional $2q^2 - 2pq + 8$ bags. If he sold $3p^2 - 2pq - 3$ bags, how many bags are still left?

5.  I went shopping with a $10 note. I bought some groceries for $2m^2 - 3mn - 2n^2$ dollars and some stationery for $3mn - 5$ dollars. How much money was I left with?

6.  A man decided to divide his $2a^3 - a^2b + 3ab^2 - 5$ hectares of land to give to his three children. If two of his children were given $a^3 - 3a^2b - 2 + ab^2$ and $4 + 2ab^2 - 7a^2b$ hectares, how many hectares did the third child get?

7.  In a farm there are $5x^3 - x^2y + 2xy^2 + y^3$ trees. Of these $3x^3 - 2x^2y - y^3 - 3xy^2$ trees are cut down. In another farm there are $2x^3 + 3xy^2 - 2y^3$ trees and another $5xy^2 - 3y^3 + 2x^2y$ trees are planted. How many trees are there now in total in both farms?

8.  In four classes of a school there are in total $5x^3 + y^3$ students. In the first three classes there are $2x^3 - x^2y + xy^2$, $2x^2y + 3xy^2$ and $3x^3 - 4xy^2 - 4 - x^2y$ students respectively. How many students are in the fourth class?

------------------------------------------------------------------------------------------------

| Solutions: Exercise 4.2 | |
|---|---|
| **1.** $13x^2 - 6xy + 3y^2$ | **5.** $15 - 2m^2 + 2n^2$ |
| **2.** $3m^2$ | **6.** $a^3 + 9a^2b - 7$ |
| **3.** $x^3 + 2x + 14$ | **7.** $4x^3 + 3x^2y + 13xy^2 - 3y^3$ |
| **4.** $p^2 + pq + 12 + q^2$ | **8.** $y^3 + 4$ |

# CHAPTER 5

Multiplication involves multiplying a term called the multiplicand by another term called the multiplier. The result is called the **product** of the two terms.

Two terms with **like signs** multiplied together give a **positive answer**.
Two terms with **unlike signs** multiplied together give a **negative answer**.

- $(+) \times (+) = (+)$
- $(+) \times (-) = (-)$

- $(-) \times (-) = (+)$
- $(-) \times (+) = (-)$

When we write a variable term x, it is assumed that its coefficient is 1. It is customary to write 1x by leaving out the coefficient. If the variable term is –x, then its coefficient is –1.

When a variable **ax** is multiplied by a constant term **b** the result is **abx**.

When multiplying two variables, the coefficients are multiplied and written in front of the product of the variables.

When **ax** is multiplied by **bx** the result is $\textbf{abx}^2$.

The variable x has a 'power' 1 in both multiplicand and the multiplier. When multiplying x by x we get $x^2$ by adding the powers of the two x's. Note that the constants a and b multiplied give ab, which is written in front of $x^2$. In general, when two or more variables are multiplied together, the 'powers' of similar variables are added together.

**Examples:**

- $9x \times 4 = 36x$
- $6x \times 4x^3 = 24x^4$

- $3x \times 10y = 30xy$
- $4xy^2 \times -3x^2y = -12x^3y^3$

When there are multiple terms in an algebraic expression and it is multiplied by a single term, the latter multiplies each term in the first expression.

- Multiply  ab – bc and ab;

$(ab - bc) \times ab = a^2b^2 - ab^2c$

- Multiply  $3x^2y - 4xy^2 + 5$ and $2xy^2$;

$(3x^2y - 4xy^2 + 5) \times 2xy^2 = 6x^3y^3 - 8x^2y^4 + 10xy^2$

**EXERCISE 5.1**----------------------------------------------------------------------------------------

Write down the product of the following:

1. $7x$ and $8$    3. $3x$ and $4x^2$    5. $7m^5$ and $9m^3$    7. $13n^3$ and $7n^7$

2. $x^2$ and $x^3$    4. $5a^3$ and $6a^4$    6. $5ax$ and $7ax$    8. $4pq$ and $8pq^2$

----------------------------------------------------------------------------------------

| Solutions: Exercise 5.1 | | | |
|---|---|---|---|
| **1.** $56x$ | **3.** $12x^3$ | **5.** $63m^8$ | **7.** $91n^{10}$ |
| **2.** $x^5$ | **4.** $30a^7$ | **6.** $35a^2x^2$ | **8.** $32p^2q^3$ |

**EXERCISE 5.2**----------------------------------------------------------------------------------------

Write down the result of the following multiplication:

1. $7bc \times 5bd$    4. $2a^2 \times 3b^3 \times 4c^2$    7. $2xy^2 \times 5yz^2 \times 3xz^3$

2. $3x^2 \times x^3y$    5. $5a^3b \times 2ab \times 7a^4b^2$    8. $6pq^3 \times p^3q \times 7pq^4$

3. $4x^2y^3 \times 3y^3$    6. $2abc \times 3bc \times 4ac$

----------------------------------------------------------------------------------------

| Solutions: Exercise 5.2 | | | |
|---|---|---|---|
| **1.** $35b^2cd$ | **3.** $12x^2y^6$ | **5.** $70a^8b^4$ | **7.** $30x^2y^3z^5$ |
| **2.** $3x^5y$ | **4.** $24a^2b^3c^2$ | **6.** $24a^2b^2c^3$ | **8.** $42p^5q^8$ |

**EXERCISE 5.3**----------------------------------------------------------------------------------------

Multiply the first algebraic expression by the second:

1.    $xy - yz$ and $xy^2$    5.    $a^2b - a^3c + 4bc^4$ and $a^2bc^2$

2.    $4m^2 - 3n^2$ and $2m^2n$    6.    $m^2n + 4mn + 7n$ and $5m^3n$

3.    $p^2 - 2q^3$ and $4r^2$    7.    $4am^2 - b^2n + 6$ and $a^3mn$

4.    $ab^2 - 5b^2c - 3$ and $7bc$    8.    $x^3 - 6a^2x^2 + 9$ and $3a^3x^2$

----------------------------------------------------------------------------------------

| Solutions: Exercise 5.3 | |
|---|---|
| **1.** $x^2y^3 - xy^2z$ | **5.** $a^4b^2c^2 - a^5bc^3 + 4a^2b^2c^6$ |
| **2.** $8m^4n - 6m^2n^3$ | **6.** $5m^5n^2 + 20m^4n^2 + 35m^3n^2$ |
| **3.** $4p^2r^2 - 8q^3r^2$ | **7.** $4a^4m^3n - a^3b^2mn^2 + 6a^3mn$ |
| **4.** $7ab^3c - 35b^3c^2 - 21bc$ | **8.** $3a^3x^5 - 18a^5x^4 + 27a^3x^2$ |

# CHAPTER 6

## ADVANCED MULTIPLICATION

In this chapter we will first look at how two simple algebraic expressions are multiplied together. We will then look at how a simple expression can be multiplied by itself a number of times and then finally look at how more difficult multiplications are done.

**Examples:**

- Multiply $a + 5$ and $a + 6$

$$
\begin{array}{l}
a + 5 \\
\underline{a + 6} \\
a^2 + 5a \qquad \text{(Multiplying } a + 5 \text{ by } a) \\
\underline{\quad + 6a + 30} \qquad \text{(Multiplying } a + 5 \text{ by } 6) \\
a^2 + 11a + 30
\end{array}
$$

The above result can be presented as $(a + 5)(a + 6) = a^2 + 11a + 30$

Observe that:

- $a \times a$ gives $a^2$    ('a' in the first expression by 'a' in the second)
- $5 \times a$ added to $a \times 6$ gives $11a$
- $5 \times 6$ gives $30$

- Multiply $5y + 3$ and $4y + 6$

$(5y + 3)(4y + 6) = 20y^2 + 42y + 18$

Observe that:

- $5y \times 4y$ gives $20y^2$,
- $3 \times 4y$ added to $5y \times 6$ gives $42y$
- $3 \times 6$ gives $18$

**More examples:**

- $(4x - 7)(6x + 1) = 24x^2 - 38x - 7$
- $(3mn - 5)(7mn - 3) = 21m^2n^2 - 44mn + 15$
- $(3x + y)(3x - y) = 9x^2 - y^2$

Note that the result of multiplying the sum of two numbers by their difference is the difference between the squares of the two numbers

16

**EXERCISE 6.1**----------------------------------------------------------------

Write down the product of:

1. $a + 4$; $a + 3$    3. $m + 11$; $m + 2$    5. $p - 7$;  $p + 6$    7. $x + 9$; $x - 11$

2. $x + 7$; $x + 8$    4. $a - 4$; $a + 3$    6. $x - 5$;  $x - 7$    8. $-y + 2$; $-y - 4$

----------------------------------------------------------------

| Solutions: Exercise 6.1 | | | |
|---|---|---|---|
| **1.** $a^2 + 7a + 12$ | **3.** $m^2 + 13m + 22$ | **5.** $p^2 - p - 42$ | **7.** $x^2 - 2x - 99$ |
| **2.** $x^2 + 15x + 56$ | **4.** $a^2 - a - 12$ | **6.** $x^2 - 12x + 35$ | **8.** $y^2 + 2y - 8$ |

**EXERCISE 6.2**----------------------------------------------------------------

Expand the following:

1. $(l - 8)(l - 4)$        4. $(m - 8)(m - 11)$        7. $(x - 7)(-x - 15)$

2. $(p - 10)(p - 7)$        5. $(q - 7)(q - 13)$        8. $(a - 12)(a - 11)$

3. $(x - 9)(x - 12)$        6. $(-q + 8)(q - 14)$

----------------------------------------------------------------

| Solutions: Exercise 6.2 | | | |
|---|---|---|---|
| **1.** $l^2 - 12l + 32$ | **3.** $x^2 - 21x + 108$ | **5.** $q^2 - 20q + 91$ | **7.** $-x^2 - 8x + 105$ |
| **2.** $p^2 - 17p + 70$ | **4.** $m^2 - 19m + 88$ | **6.** $-q^2 + 22q - 112$ | **8.** $a^2 - 23a + 132$ |

**EXERCISE 6.3**----------------------------------------------------------------

Write down the product of:

1.    $2a + 1$;  $a + 2$        5.    $3x - 2y$;  $2x + 3y$

2.    $2x - 1$;  $x - 3$        6.    $5p + 4q$;  $5p - 4q$

3.    $4x - 2$;  $x - 7$        7.    $a - 3x$;  $3a - 2x$

4.    $2m - 3n$;  $3m - 4n$        8.    $4b + c$;  $6b - 5c$

----------------------------------------------------------------

| Solutions: Exercise 6.3 | | | |
|---|---|---|---|
| **1.** $2a^2 + 5a + 2$ | **3.** $4x^2 - 30x + 14$ | **5.** $6x^2 + 5xy - 6y^2$ | **7.** $3a^2 - 11ax + 6x^2$ |
| **2.** $2x^2 - 7x + 3$ | **4.** $6m^2 - 17mn + 12n^2$ | **6.** $25p^2 - 16q^2$ | **8.** $24b^2 - 14bc - 5c^2$ |

**EXERCISE 6.4**----------------------------------------------------

1. $(3m - 5n)(5m + 4n)$

2. $(7x - 3y)(3x - 5y)$

3. $(6k - 7l)(3k + 4l)$

4. $(9n - 5l)(9n + 5l)$

5. $(12a - b)(a - 8b)$

6. $(11m - 2n)(m + 7n)$

7. $(x + 9y)(13x - 7y)$

8. $(5x + 9y)(3x - 11y)$

------------------------------------------------------------

**Solutions: Exercise 6.4**

| | |
|---|---|
| **1.** $15m^2 - 13mn - 20n^2$ | **5.** $12a^2 - 97ab + 8b^2$ |
| **2.** $21x^2 - 44xy + 15y^2$ | **6.** $11m^2 + 75mn - 14n^2$ |
| **3.** $18k^2 + 3kl - 28l^2$ | **7.** $13x^2 + 110xy - 63y^2$ |
| **4.** $81n^2 - 25l^2$ | **8.** $15x^2 - 28xy - 99y^2$ |

**EXERCISE 6.5**----------------------------------------------

Expand the following:

1. $(ab + c)(2ab + 3c)$

2. $(xy + 2a)(xy - 2a)$

3. $(pq - m)(2pq + m)$

4. $(2 - 3abc)(3 - 2abc)$

5. $(1 + 6b)(1 - 7a)$

6. $(3 + ab)(4 - ab)$

7. $(x^2 + y^2)(x^2 - y^2)$

8. $(2a^2 + b^2)(2a^2 - b^2)$

------------------------------------------------------------

**Solutions: Exercise 6.5**

| | |
|---|---|
| **1.** $2a^2b^2 + 5abc + 3c^2$ | **5.** $1 + 6b - 7a - 42ab$ |
| **2.** $x^2y^2 - 4a^2$ | **6.** $12 + ab - a^2b^2$ |
| **3.** $2p^2q^2 - mpq - m^2$ | **7.** $x^4 - y^4$ |
| **4.** $6 - 13abc + 6a^2b^2c^2$ | **8.** $4a^4 - b^4$ |

**EXERCISE 6.6**----------------------------------------------

Expand the following:

1. $(a^2 + 5b)(a^2 - 7b)$

2. $(2x^2 - y)(3x - 1)$

3. $(5m^2 - 2n^3)(m^2 + n^3)$

4. $(m^4 + n)(m^4 - n)$

5. $(2x^2 - y^2)(x^2 - y^2)$

6. $(p^3q - r)(p^3q + r)$

7. $(m^2n - 3)(m^2n - 5)$

8. $(p^3q - 8)(p^3q - 3)$

------------------------------------------------------------

| Solutions: Exercise 6.6 | |
|---|---|
| **1.** $a^4 - 2a^2b - 35b^2$ | **5.** $2x^4 - 3x^2y^2 + y^4$ |
| **2.** $6x^3 - 3xy - 2x^2 + y$ | **6.** $p^6q^2 - r^2$ |
| **3.** $5m^4 + 3m^2n^3 - 2n^6$ | **7.** $m^4n^2 - 8m^2n + 15$ |
| **4.** $m^8 - n^2$ | **8.** $p^6q^2 - 11p^3q + 24$ |

## MULTIPLYING AN ALGEBRAIC EXPRESSION BY ITSELF

**Examples:**

- Expand $(a + b)^2$

$(a + b)^2 = a^2 + 2ab + b^2$

Observe that:

- the first term is $a^2$
- the second term is $2ab$    (a's power is decreased by 1 and b introduced)
- the third term is $b^2$    (a is dropped and b's power is increased by 1)

The coefficient of $a^2$ is 1, the coefficient of $ab$ is 2 and the coefficient of $b^2$ is 1

Since there are **three terms**, the coefficients can be determined by the numbers in the **third row** of Pascal's Triangle shown below:

<u>**Pascal's Triangle**</u>

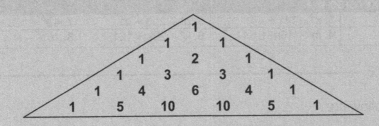

- Expand $(a - b)^2$

$(a - b)^2 = a^2 - 2ab + b^2$    (obtained by multiplying in the usual way)

Observe that the coefficients are the same as in the case of $(a + b)^2$ except that the middle term has a **negative sign**. As a rule, **the signs of the coefficients will alternate,** starting from a positive sign for the first term when we expand an expression of the form $(a - b)^n$.

- Expand $(a - b)^3$

$(a - b)^3 = a^3 - 3a^2b + 3ab^2 - b^3$

Observe how the powers of 'a' decrease and the powers of 'b' increase as we move from term to term. Also observe that the coefficients are obtained from the fourth row of the Pascal's Triangle and that the signs alternate starting from a positive sign for the first term.

- Expand $(a + b)^4$

$$(a + b)^4 = a^4 + 4a^3b + 6a^2b^2 + 4ab^3 + b^4$$

- Expand $(x - 2y)^4$

$$(x - 2y)^4 = x^4 - 4x^3(2y) + 6x^2(2y)^2 - 4x(2y)^3 + (2y)^4$$

$$= x^4 - 8x^3y + 24x^2y^2 - 32xy^3 + 16y^4$$

## EXERCISE 6.7

Expand the following:

| | | | |
|---|---|---|---|
| 1. $(x + 2)^2$ | 3. $(x - 7)^2$ | 5. $(p - q)^2$ | 7. $(2a + 3)^2$ |
| 2. $(a + 5)^2$ | 4. $(m - 8)^2$ | 6. $(k + 11)^2$ | 8. $(4a - 5)^2$ |

### Solutions: Exercise 6.7

| | | | |
|---|---|---|---|
| 1. $x^2 + 4x + 4$ | 3. $x^2 - 14x + 49$ | 5. $p^2 - 2pq + q^2$ | 7. $4a^2 + 12a + 9$ |
| 2. $a^2 + 10a + 25$ | 4. $m^2 - 16m + 64$ | 6. $k^2 + 22k + 121$ | 8. $16a^2 - 40a + 25$ |

## EXERCISE 6.8

Expand the following:

| | | | |
|---|---|---|---|
| 1. $(3m + 8)^2$ | 3. $(2a - b)^2$ | 5. $(8x + y)^2$ | 7. $(2m - 7n)^2$ |
| 2. $(2p - 7)^2$ | 4. $(p + 2q)^2$ | 6. $(4x - 3y)^2$ | 8. $(5a - 10b)^2$ |

### Solutions: Exercise 6.8

| | |
|---|---|
| 1. $9m^2 + 48m + 64$ | 5. $64x^2 + 16xy + y^2$ |
| 2. $4p^2 - 28p + 49$ | 6. $16x^2 - 24xy + 9y^2$ |
| 3. $4a^2 - 4ab + b^2$ | 7. $4m^2 - 28mn + 48n^2$ |
| 4. $p^2 + 4pq + 4q^2$ | 8. $25a^2 - 100ab + 100b^2$ |

Expand the following:

1. $(3a + 2b)^2$     3. $(pq + 2)^2$     5. $(2a^2 + 5)^2$     7. $(mn^2 - 3)^2$

2. $(ab - c)^2$     4. $(a^2 - b^2)^2$     6. $(x^2 - 2y^2)^2$     8. $(3xy + 5yz)^2$

---

| Solutions: Exercise 6.9 | |
|---|---|
| **1.** $9a^2 + 12ab + 4b^2$ | **5.** $4a^4 + 20a^2 + 25$ |
| **2.** $a^2b^2 - 2abc + c^2$ | **6.** $x^4 - 4x^2y^2 + 4y^4$ |
| **3.** $p^2q^2 + 4pq + 4$ | **7.** $m^2n^4 - 6mn^2 + 9$ |
| **4.** $a^4 - 2a^2b^2 + b^4$ | **8.** $9x^2y^2 + 30xy^2z + 25y^2z^2$ |

Expand using the Pascal's Triangle:

1. $(x + y)^3$     3. $(x + y)^4$     5. $(p + 3q)^4$     7. $(3 + 5y)^4$

2. $(3p - 2)^3$     4. $(a - 2b)^4$     6. $(a + b)^5$     8. $(2m - 3n)^5$

---

| Solutions: Exercise 6.10 |
|---|
| **1.** $x^3 + 3x^2y + 3xy^2 + y^3$ |
| **2.** $27p^3 - 54p^2 + 36p - 8$ |
| **3.** $x^4 + 4x^3y + 6x^2y^2 + 4xy^3 + y^4$ |
| **4.** $a^4 - 8a^3b + 24a^2b^2 - 32ab^3 + 16b^4$ |
| **5.** $p^4 + 12p^3q + 54p^2q^2 + 108pq^3 + 81q^4$ |
| **6.** $a^5 + 5a^4b + 10a^3b^2 + 10a^2b^3 + 5ab^4 + b^5$ |
| **7.** $81 + 540y + 1350y^2 + 1500y^3 + 625y^4$ |
| **8.** $32m^5 - 240m^4n + 720m^3n^2 - 1080m^2n^3 + 810mn^4 - 243n^5$ |

## SQUARING EXPRESSIONS WITH MORE THAN TWO TERMS

The result of squaring an expression with more than two terms is obtained by adding the sum of the squares of each term to twice the product of two terms taken at a time in a cyclic order.

**Examples:**

- $(a + b + c)^2 = a^2 + b^2 + c^2 + 2ab + 2bc + 2ca$     (Cyclic order is: ab, bc, ca)

- $(a - b - c)^2 = a^2 + b^2 + c^2 - 2ab + 2bc - 2ca$     (Note the changes in sign)

- $(a + b + c + d)^2 = a^2 + b^2 + c^2 + d^2 + 2ab + 2bc + 2cd + 2da + 2ac + 2bd$

- $(a - b - c - d)^2 = a^2 + b^2 + c^2 + d^2 - 2ab + 2bc + 2cd - 2da - 2ac + 2bd$

**EXERCISE 6.11**-------------------------------------------------------------------------------------------------

Square the following algebraic expressions:

| | | | |
|---|---|---|---|
| 1. $x + y + z$ | 3. $p + q - r$ | 5. $x - y - z$ | 7. $2a - b + 3c$ |
| 2. $x - y + z$ | 4. $l + m + n$ | 6. $2l - m + n$ | 8. $x - 3y - z$ |

-------------------------------------------------------------------------------------------------

| **Solutions: Exercise 6.11** | |
|---|---|
| **1.** $x^2 + y^2 + z^2 + 2xy + 2yz + 2zx$ | **5.** $x^2 + y^2 + z^2 - 2xy + 2yz - 2zx$ |
| **2.** $x^2 + y^2 + z^2 - 2xy - 2yz + 2zx$ | **6.** $4l^2 + m^2 + n^2 - 4lm - 2mn + 4nl$ |
| **3.** $p^2 + q^2 + r^2 + 2pq - 2qr - 2rp$ | **7.** $4a^2 + b^2 + 9c^2 - 4ab + 12ac - 6bc$ |
| **4.** $l^2 + m^2 + n^2 + 2lm + 2mn + 2nl$ | **8.** $x^2 + 9y^2 + z^2 - 6xy - 2xz + 6yz$ |

**EXERCISE 6.12**-------------------------------------------------------------------------------------------------

Square the following algebraic expressions:

| | | |
|---|---|---|
| 1. $2x + y + z$ | 4. $5p - q + 2$ | 7. $2a - 3b - c + d$ |
| 2. $1 + x + 2y$ | 5. $a + b + c + d$ | 8. $p - 2q + 3r + s$ |
| 3. $2p - 3q - 1$ | 6. $a - b + c - d$ | |

-------------------------------------------------------------------------------------------------

| **Solutions: Exercise 6.12** | |
|---|---|
| **1.** $4x^2 + y^2 + z^2 + 4xy + 2yz + 4xz$ | **5.** $a^2 + b^2 + c^2 + d^2 + 2ab + 2ac + 2ad + 2bc + 2bd + 2cd$ |
| **2.** $x^2 + 4y^2 + 4xy + 2x + 4y + 1$ | **6.** $a^2 + b^2 + c^2 + d^2 - 2ab + 2ac - 2ad - 2bc + 2bd - 2cd$ |
| **3.** $4p^2 + 9q^2 - 12pq - 4p + 6q + 1$ | **7.** $4a^2 + 9b^2 + c^2 + d^2 - 12ab - 4ac + 4ad + 6bc - 6bd - 2cd$ |
| **4.** $25p^2 + q^2 - 10pq + 20p - 4q + 4$ | **8.** $p^2 + 4q^2 + 9r^2 + s^2 - 4pq + 6pr + 2ps - 12qr - 4qs + 6rs$ |

## THE PRODUCT OF ANY TWO ALGEBRAIC EXPRESSIONS

In order to find the product of any two expressions we carry out long multiplication as set out in the following examples:

**Examples:**

- Find the product of $3x^2 + 2x - 5$ and $x - 3$

| | | | |
|---|---|---|---|
| $3x^2$ | $+2x$ | $-5$ | (Multiplicand) |
| | $x$ | $-3$ | (Multiplier) |
| $-9x^2$ | $-6x$ | $+15$ | (Multiplying the multiplicand by '$-3$') |
| $3x^3$ $+2x^2$ | $-5x$ | | (Multiplying the multiplicand by '$x$') |
| $3x^3$ $-7x^2$ | $-11x$ | $+15$ | |

Note how the like terms are written one below the other for easy addition.

- Find the product of $2a^2 - 3a + 2$ and $a^2 + 2a - 1$

$$
\begin{array}{rrrrr}
 & 2a^2 & -3a & 2 & \\
 & a^2 & +2a & -1 & \\
\hline
 & -2a^2 & +3a & -2 & \\
4a^3 & -6a^2 & +4a & & \\
2a^4 & -3a^3 & +2a^2 & & \\
\hline
2a^4 & +a^3 & -6a^2 & +7a & -2 \\
\end{array}
$$

## EXERCISE 6.13----------------------------------------------------------------------------

Find the product of each of the following pairs of algebraic expressions:

1. $x^2 - x - 2$;  $x - 4$

2. $a^2 + 2a + 4$;  $a - 2$

3. $2a^2 - 3a + 1$;  $2a - 1$

4. $3x^2 + 5x + 4$;  $3x - 5$

5. $2m^2 - 3m - 1$;  $3m^2 - 1$

6. $4p^2 - 2p + 7$;  $3p - 4$

7. $c^2 - 4c + 7$;  $2c - 4$

8. $(x + 2y)^2$;  $3x - 4$

---

| Solutions: Exercise 6.13 | |
|---|---|
| 1. $x^3 - 5x^2 + 2x + 8$ | 5. $6m^4 - 9m^3 - 5m^2 + 3m + 1$ |
| 2. $a^3 - 8$ | 6. $12p^3 - 22p^2 + 29p - 28$ |
| 3. $4a^3 - 8a^2 + 5a - 1$ | 7. $2c^3 - 12c^2 + 30c - 28$ |
| 4. $9x^3 - 13x - 20$ | 8. $3x^3 + 12x^2y + 12xy^2 - 4x^2 - 16xy - 16y^2$ |

## EXERCISE 6.14----------------------------------------------------------------------------

Find the product of each of the following pairs of algebraic expressions:

1. $a^2 - ab + b^2$;  $a^2 + ab + b^2$

2. $x^3 - 5x + 6$;  $x^3 + 5x - 5$

3. $x^2 + 2x + 4$;  $x^3 - x + 2$

4. $a^3 - 3a + 2$;  $a^2 + 3a - 2$

5. $x^3 + y^3 - x^2y^2$;  $x^2y^2 - x^3 + y^3$

6. $x^2 + y^2 + z^2 + xy + yz - zx$;  $x - y - z$

7. $(x - y)^3$;  $(x + y)$

8. $(2x + y)^3$;  $(2x - y)$

| **Solutions: Exercise 6.14** |
|---|
| **1.** $a^4 + a^2b^2 + b^4$ |
| **2.** $x^6 + x^3 - 25x^2 + 55x - 30$ |
| **3.** $x^5 + 2x^4 + 3x^3 + 8$ |
| **4.** $a^5 + 3a^4 - 5a^3 - 7a^2 + 12a - 4$ |
| **5.** $2x^5y^2 - x^6 + y^6 - x^4y^4$ |
| **6.** $x^3 + 2xz^2 + xyz - 2zx^2 - y^3 - 2yz^2 - 2y^2z - z^3$ |
| **7.** $x^4 - 2x^3y + 2xy^3 - y^4$ |
| **8.** $16x^4 + 16x^3y - 4xy^3 - y^4$ |

## EXERCISE 6.15----------------------------------------------------------------------------

1. If wages are paid at the rate of $3x + 8$ dollars per hour, what will be the total wages for $9x - 13$ hours?

2. If a box of books weighs $2x^2 + 3y^2$ kilograms, how much will $3x^2 + 4y^2$ of identical boxes weigh?

3. The price of a chair is $x - y$ dollars. A table is $x^2 + xy + y^2$ times the price of a chair. What is the price of the table?

4. There are a number of light posts in a straight line along a street. If the distance between any two adjacent light posts is $9p^2 - 4q^2$ metres, what is the distance between $3p^3 - 2q^2 + 1$ light posts?

5. If a bottle of coffee costs $4x - 5y$ dollars, how much will $x^2 - 3x - y^2$ bottles cost?

6. If a train travels at $1 - x^2 + 2y^2$ kilometres per hour, what distance will it cover in $x^2 + y^2$ hours?

7. What would be the result of multiplying the square of $3x - 2y$ by $x - y$?

8. Determine the product of the cube of $a - 3b$ and $2abc$.

-----------------------------------------------------------------------------------------------

| **Solutions: Exercise 6.15** | |
|---|---|
| **1.** $\$(27x^2 + 33x - 104)$ | **5.** $\$(4x^3 - 12x^2 - 5x^2y + 15xy - 4xy^2 + 5y^3)$ |
| **2.** $(6x^4 + 17x^2y^2 + 12y^4)$ kg | **6.** $(-x^4 + x^2 + x^2y^2 + 2y^4 + y^2)$ km |
| **3.** $\$(x^3 - y^3)$ | **7.** $9x^3 - 21x^2y + 16xy^2 - 4y^3$ |
| **4.** $(27p^5 - 12p^3q^2 - 18p^2q^2 + 8q^4)$ metres | **8.** $2a^4bc - 18a^3b^2c + 54a^2b^3c - 54ab^4c$ |

Division involves dividing a term called the **dividend** by another term called the **divisor.** The result of the division is called the **quotient**. Some divisions result in a **remainder** if the divisor cannot divide the dividend exactly.

---

If both the dividend and divisor are terms with **like signs** the quotient will be **positive**. When the terms are with **unlike signs**, a **negative** answer results:

- $(+) \div (+) = (+)$
- $(+) \div (-) = (-)$

- $(-) \div (-) = (+)$
- $(-) \div (+) = (-)$

---

When a variable **ax** is divided by a constant term **b** the result is $\dfrac{a}{b}x$

---

When we divide $15x^3$ by $3x$ the result is $5x^2$. The variable x has a 'power' 3 in the dividend and 1 in the divisor. We get $x^2$ by subtracting the power of x in the divisor from the power of x in the dividend. Note that the coefficient of $x^3$ is divided by the coefficient of x and the answer 5 is written in front of $x^2$.

In general, when two or more of the same variables are involved in the division, the 'power' of the variable in the denominator is subtracted from the power of the variable in the numerator.

---

**Examples:**

- $12x \div 4 = 3x$
- $-24m \div 8 = -3m$
- $-24a \div 8 = -3a$
- $-20a \div -4x = 5$

- $18y^6z \div 3y^2 = 6y^4z$
- $x^4y^3 \div -x^2y = -x^2y^2$
- $5y^3z^4 \div y^2z^5 = 5yz^{-1}$

---

**EXERCISE 7.1** -------------------------------------------------------------

Carry out the following simple division:

1. $8x \div 4$

3. $-18m \div 6$

5. $4ab \div 2b$

7. $-32m^2 \div 8m$

2. $12a \div 3$

4. $-21p \div -7$

6. $6x^2 \div 2x$

8. $-42p^2 \div 6p$

------------------------------------------------------------------------------

| **Solutions: Exercise 7.1** | | | | | | | |
|---|---|---|---|---|---|---|---|
| **1.** 2x | **2.** 4a | **3.** −3m | **4.** 3p | **5.** 2a | **6.** 3x | **7.** −4m | **8.** −7p |

Divide the first term by the second:

1. $27a^3b$;  $9ab$

2. $18x^2y$;  $-9xy$

3. $54p^3q^3r$;  $9q^2r$

4. $-30m^5n^2$;  $-6m^2n^2$

5. $x^2y^2$;  $-2xy$

6. $-35a^{10}b^4$;  $-7a^6b^3$

7. $12x^4y^3z^6$;  $3x^2yz^3$

8. $25a^3b^2c^2$;  $5abc$

| Solutions: Exercise 7.2 | | | | | | | |
|---|---|---|---|---|---|---|---|
| 1. $3a^2$ | 2. $-2x$ | 3. $6p^3q$ | 4. $5m^3$ | 5. $\frac{-xy}{2}$ | 6. $5a^4b$ | 7. $4x^2y^2z^3$ | 8. $5a^2bc$ |

## DIVIDEND AS AN ALGEBRAIC EXPRESSION

When the divisor is a simple algebraic expression, each term in the dividend is divided by the divisor.

- Divide:  $6x^3y^4 + 9x^4y^5$ by $3xy^2 = 2x^2y^2 + 3x^3y^3$

- Simplify:  $\dfrac{-24m^2n^2 + 18m^3}{-6m^2n^2} = 4 - 3mn^{-2}$

Divide:

1.  $48a^2b - 8ab^2$ by $8ab$

2.  $5x^3y - 7x^2y^2$ by $x^2y$

3.  $3p^3 - 6p^2q + 12pq^2$ by $-3p$

4.  $5y^3 + 35ay^2 - 20y$ by $-5y$

Simplify:

5.  $\dfrac{25a^2b^3c^3 - 15ab^2c^2}{5abc}$

6.  $\dfrac{27x^3y^2z^2 - 18x^5y^3z^3}{9x^2yz^2}$

7.  $\dfrac{18l^2m^3n^6 - 10l^3m^2n^4}{2l^2m^2n^4}$

8.  $\dfrac{121x^4y^3z^7 - 55x^3y^5z^4}{11x^2y^2z^3}$

| Solutions: Exercise 7.3 | | | |
|---|---|---|---|
| 1. $6a - b$ | 3. $-p^2 + 2pq - 4q^2$ | 5. $5ab^2c^2 - 3bc$ | 7. $9mn^2 - 5l$ |
| 2. $5x - 7y$ | 4. $-y^2 - 7ay + 4$ | 6. $3xy - 2x^3y^2z$ | 8. $11x^2yz^4 - 5xy^3z$ |

When dividing one algebraic expression (dividend) by another algebraic expression (the divisor) the order of the terms is important. We **arrange the terms in both the dividend and the divisor in a descending order.**

If the divisor does not divide the dividend exactly, the amount left over is called the **remainder.**

**Example 1:**

Divide $-2x^3 + x^4 - 9x + 5x^2 - 2$ by $x - 2$

- Arrange the dividend and divisor in descending order of the powers of x:

$$x - 2 \overline{\smash{\big)}\ x^4 \quad -2x^3 \quad +5x^2 \quad -9x \quad -2}$$

- Ask yourself what term would multiply x in the divisor $x - 2$ to give $x^4$. The answer is $x^3$:

$$
\begin{array}{r}
x^3 \phantom{aaaaaaaaaaaaaaaaa} \\
x - 2 \overline{\smash{\big)}\ x^4 \quad -2x^3 \quad +5x^2 \quad -9x \quad -2}
\end{array}
$$

- Multiply $x - 2$ by $x^3$ after writing $x^3$ on top of $x^4$, and write down the result as shown:

$$
\begin{array}{r}
x^3 \phantom{aaaaaaaaaaaaaaaaa} \\
x - 2 \overline{\smash{\big)}\ x^4 \quad -2x^3 \quad +5x^2 \quad -9x \quad -2} \\
x^4 \quad -2x^3 \phantom{aaaaaaaaaaaaa}
\end{array}
$$

- Subtract and bring down the next two terms $5x^2 - 9x$:

$$
\begin{array}{r}
x^3 \phantom{aaaaaaaaaaaaaaaaa} \\
x - 2 \overline{\smash{\big)}\ x^4 \quad -2x^3 \quad +5x^2 \quad -9x \quad -2} \\
\underline{x^4 \quad -2x^3} \phantom{aaaaaaaaaaaaa} \\
0 \quad +0 \quad 5x^2 \quad -9x \phantom{aa}
\end{array}
$$

- Ask yourself what term would multiply x to give $+5x^2$. The answer is $5x$:

$$
\begin{array}{r}
x^3 \phantom{aaaaa} +5x \phantom{aaaaaaa} \\
x - 2 \overline{\smash{\big)}\ x^4 \quad -2x^3 \quad +5x^2 \quad -9x \quad -2} \\
\underline{x^4 \quad -2x^3} \phantom{aaaaaaaaaaaaa} \\
5x^2 \quad -9x \phantom{aa}
\end{array}
$$

- Multiply x – 2 by 5x and write the result as shown:

$$
\begin{array}{r}
x^3 \qquad\quad +5x \\
x-2\ \overline{\big)\ x^4 \quad -2x^3 \quad +5x^2 \quad -9x \quad -2} \\
\underline{x^4 \quad -2x^3} \\
5x^2 \quad -9x \\
\underline{\mathbf{5x^2 \quad -10x}}
\end{array}
$$

- Subtract to get x, and then bring down –2:

$$
\begin{array}{r}
x^3 \qquad\quad +5x \\
x-2\ \overline{\big)\ x^4 \quad -2x^3 \quad +5x^2 \quad -9x \quad -2} \\
\underline{x^4 \quad -2x^3} \\
5x^2 \quad -9x \\
\underline{5x^2 \quad -10x} \\
\mathbf{x \quad -2}
\end{array}
$$

- Ask yourself what term would multiply x to give x.  The answer is 1:

$$
\begin{array}{r}
x^3 \qquad\quad +5x \qquad \mathbf{+1} \\
x-2\ \overline{\big)\ x^4 \quad -2x^3 \quad +5x^2 \quad -9x \quad -2} \\
\underline{x^4 \quad -2x^3} \\
5x^2 \quad -9x \\
\underline{5x^2 \quad -10x} \\
x \quad -2
\end{array}
$$

- Multiply x – 2 by 1 and write the result as shown:

$$
\begin{array}{r}
x^3 \qquad\quad +5x \qquad +1 \\
x-2\ \overline{\big)\ x^4 \quad -2x^3 \quad +5x^2 \quad -9x \quad -2} \\
\underline{x^4 \quad -2x^3} \\
5x^2 \quad -9x \\
\underline{5x^2 \quad -10x} \\
x \quad -2 \\
\underline{\mathbf{x \quad -2}}
\end{array}
$$

- Subtract to get zero.  The division has ended:

$$
\begin{array}{r}
x^3 \qquad\quad +5x \qquad +1 \\
x-2\ \overline{\big)\ x^4 \quad -2x^3 \quad +5x^2 \quad -9x \quad -2} \\
\underline{x^4 \quad -2x^3} \\
5x^2 \quad -9x \\
\underline{5x^2 \quad -10x} \\
x \quad -2 \\
\underline{x \quad -2} \\
\mathbf{0 \quad +0}
\end{array}
$$

The answer (or quotient) for the division is:  $x^3 + 5x + 1$ with no remainder.

- Divide $y^4 - 3y^3 - y^2 + 9y - 9$ by $y^2 - 3$

$$
\begin{array}{r}
y^2 \quad -3y \quad +2 \\
y^2 - 3 \, \overline{\smash{)}\, y^4 \quad -3y^3 \quad -y^2 \quad +9y \quad -9} \\
\underline{y^4 \qquad\qquad -3y^2} \\
-3y^3 \quad +2y^2 \quad +9y \\
\underline{-3y^3 \qquad\qquad +9y} \\
+2y^2 \quad +0 \quad -9 \\
\underline{+2y^2 \qquad\quad -6} \\
-3
\end{array}
$$

The answer for the division is: Quotient $y^2 - 3y + 2$ and remainder $-3$.

## EXERCISE 8.1 --------------------------------------------------------------

Divide:

1. $x^2 + 3x + 2$ by $x + 1$

2. $a^2 + 13a + 42$ by $a + 7$

3. $m^2 + 6m - 16$ by $m - 2$

4. $p^2 + 4p - 21$ by $p - 3$

5. $p^2 - 20p + 99$ by $p - 9$

6. $15x^2 + 14x - 8$ by $5x - 2$

7. $11 - 79x + 14x^2$ by $-1 + 7x$

8. $-91 + 2a + 8a^2$ by $7 + 2a$

### Solutions: Exercise 8.1

| **1.** $x + 2$ | **2.** $a + 6$ | **3.** $m + 8$ | **4.** $p + 7$ | **5.** $p - 11$ | **6.** $3x + 4$ | **7.** $2x - 11$ | **8.** $4a - 13$ |
|---|---|---|---|---|---|---|---|

## EXERCISE 8.2 --------------------------------------------------------------

Divide:

1. $x^2 - 9$ by $x + 3$

2. $-x^2 + 64$ by $x - 8$

3. $81m^2 - 4n^2$ by $9m + 2n$

4. $16p^2 - 49q^2$ by $4p - 7q$

5. $x^3 - 4x^2 - x + 10$ by $x - 2$

6. $11a^2 + 2a^3 - 12a + 3$ by $2a - 1$

7. $8a^3 - b^3$ by $2a - b$

8. $27x^3 - y^3$ by $3x - y$

### Solutions: Exercise 8.2

| **1.** $x - 3$ | **3.** $9m - 2n$ | **5.** $x^2 - 2x - 5$ | **7.** $4a^2 + 2ab + b^2$ |
|---|---|---|---|
| **2.** $-x - 8$ | **4.** $4p + 7q$ | **6.** $a^2 + 6a - 3$ | **8.** $9x^2 + 3xy + y^2$ |

**EXERCISE 8.3**----------------------------------------------------------------

Divide:

1. $x^4 - 3x^3 - 6x - 4$ by $x^2 + 2$

2. $6 - 9m - 42m^2$ by $2 - 7m$

3. $6p^3 - p + 7p^2 - 2$ by $3p + 2$

4. $x^4 - 7x^3 + 12x^2 - 4x$ by $x^2 - 2x$

5. $3a^3 + 8ab^2 - 4b^3 - 5a^2b$ by $3a - 2b$

6. $3a^3 + 17ax^2 - 5x^3 - 15a^2x$ by $a - x$

7. $1 - 4x^2 - 2x - 2x^4 + 7x^3$ by $1 + x - 2x^2$

8. $35 - 5y - 15y^2 - 7y^3 + y^4 + 3y^5$ by $y^3 - 5$

-------------------------------------------------------------------------

| Solutions: Exercise 8.3 | | | |
|---|---|---|---|
| **1.** $x^2 - 3x - 2$ | **3.** $2p^2 + p - 1$ | **5.** $a^2 - ab + 2b^2$ | **7.** $1 - 3x + x^2$ |
| **2.** $3 + 6m$ | **4.** $x^2 - 5x + 2$ | **6.** $3a^2 - 12ax + 5x^2$ | **8.** $3y^2 + y - 7$ |

**EXERCISE 8.4**----------------------------------------------------------------

1. If $\$(6m^5n^3 + m^3n^2)$ is divided equally between $m^3n$ children, how much will each child receive?

2. If John's weight is $3ay^2$ times that of Peter's and if John weighed $(3a^2y^3 - 12ay^4)$ kilograms, how much does Peter weigh?

3. If one revolution of a circular wheel traverses $3a - b$ cm, how many revolutions would it make when traversing a distance of $6a^2 - 5ab + b^2$ cm?

4. How long will it take for a train travelling at a constant speed of $4x - 2$ km/h to travel a distance of $32x^2 + 12x - 14$ kilometres?

5. The product of two numbers is $a^3 - 8a^2 + 23a - 40$. If one number is $a - 5$, what is the other number?

6. A piece of cloth costs $\$(b - 2)$ per metre. How many metres of the cloth can be bought for $\$(25b - 23b^2 + 6b^3 - 6)$?

7. A box can contain $(a + 1)$ mangoes. How many similar boxes would be needed to pack $(2a^3 + 3a^2 - 1)$ mangoes?

8. The area of a rectangular room is $8m^3 - 27n^3$ square metres. If the width of the room is $2m - 3n$ metres, what is its length?

-------------------------------------------------------------------------

| Solutions: Exercise 8.4 | |
|---|---|
| **1.** $\$(6m^2n^2 + n)$ | **5.** $a^2 - 3a + 8$ |
| **2.** $(ay - 4y^2)$kg | **6.** $(6b^2 - 11b + 3)$metres |
| **3.** $(2a - b)$rev | **7.** $(2a^2 + a - 1)$boxes |
| **4.** $(8x + 7)$hours | **8.** $(4m^2 + 6mn + 9n^2)$metres |

When multiplication and division are the only operations involved in the same problem it is appropriate to do the calculations in the given order of the signs. Performing a multiplication before division will give a different result to that of performing a division before multiplication.

---

**Example 1:**

Calculate: $12xy \div 3y \times 2x$

- If we do the division first: $12xy \div 3y = 4x$ and $4x \times 2x = 8x^2$

- If we do the multiplication first: $3y \times 2x = 6xy$ and $12xy \div 6xy = 2$

Note that there are two different answers. This is why we do the calculations in the given order.

**Example 2:**

Evaluate: $(x^2 - 25) \div (x + 5) \times (2x + 1)$

- Using long division, divide $(x^2 - 25)$ by $(x + 5)$ to get $x - 5$

- When $x - 5$ is multiplied by $(2x + 1)$, the final result is $2x^2 - 9x - 5$

**Example 3:**

Simplify $(a + b)^2 \times (a - b)^2 \div (a + b)(a - b)$

There is a simpler method for solving this problem that will be introduced later. At this stage it is better to solve this expression using long division and multiplication:

- First expand $(a + b)^2$ and $(a - b)^2$ to give $a^2 + 2ab + b^2$ and $a^2 - 2ab + b^2$ respectively.

- Multiplying them together gives the result $a^4 - 2a^2b^2 + b^4$

- Multiplying $(a + b)$ by $(a - b)$ gives $a^2 - b^2$

- Now we divide $a^4 - 2a^2b^2 + b^4$ by $a^2 - b^2$ to get $a^2 - b^2$

---

**EXERCISE 9.1**----------------------------------------------------------------------------------------

Evaluate using the given order of the signs:

1. $9x \times 4y \div 12y$

2. $14pq \times 9p^2 \div 6p^2q$

3. $51m^2n^2 \div 3m^2n \times 4mn^3$

4. $65a^3b^3 \div 13a^2b^4 \times 3a^3b^4$

5. $(4x^2 - 9x + 5) \div (x - 1) \times (3x + 5)$

6. $(a^2 - 16) \div (a + 4) \times (2a - 1)$

7. $(2x^3 - 15x^2 + 31x - 12) \div (2x - 1) \div (x - 4)$

8. $(27 - x^3) \div (3 - x) \times (1 - x)$

----------------------------------------------------------------------------------------

| Solutions: Exercise 9.1 | | | |
|---|---|---|---|
| **1.** $3x$ | **3.** $68mn^4$ | **5.** $12x^2 + 5x - 25$ | **7.** $x - 3$ |
| **2.** $21p$ | **4.** $15a^4b^3$ | **6.** $2a^2 - 9a + 4$ | **8.** $-x^3 - 2x^2 - 6x + 9$ |

**EXERCISE 9.2**----------------------------------------------------------------------------------------

Evaluate using the given order of the signs:

1. $(x^3 + 8y^3) \div (x + 2y) \times (x - y)$

2. $(m + n) \times (m^2 - mn + n^2) \times (m^3 - n^3)$

3. $(4 - 9x^2y^2) \div (2 - 3xy) \times (1 - xy)$

4. $(21m^2 - 19mn + 4n^2) \times (m - n) \div (7m - 4n)$

5. $(p + q)^2 \times (p - q)^2 \div (p + q)(p - q)$

6. $(a + b + c) \times (a - b - c) \times (a^2 + b^2 + 2bc + c^2)$

7. $(5x^2 - 20y^2) \div (x + 2y) \times (2x - 3y)$

8. $(x^4 - 16y^4) \div (x + 2y) \div (x^2 + 4y^2)$

----------------------------------------------------------------------------------------

## EXERCISE 9.3-------------------------------------------------------------------

1. A man walking at a steady speed covers a distance of 'b' kilometres in 'a' hours. What distance would he cover in 80 minutes?

2. A train travels at a constant speed of $38x^3y$ km/h for $25xy^5$ hours. How long will it take for the train to cover the same distance travelling at $95x^3y^2$ km/h?

3. Paying $4x - 3$ dollars per kilogram, a trader buys $4x + 3$ kilograms containing $16x^3 - 16x^2 - 9x + 9$ of chocolates. How many chocolates is he buying for one dollar?

4. $(x - y)$ bags of oranges cost $2x^2 + xy - 3y^2$ dollars. How much would $(x + y)$ bags of oranges cost?

5. There are $(5p - 3)$ bags containing $(3p - 7)$ kilograms of sugar each. If the sugar is repacked into bags each holding $(15p + 1)$ kilograms, how much sugar will remain?

6. If a man walks $(p^2 - 25q^2)$ kilometres at $(p - 5q)$ kilometres per hour, how long would he take? If he travelled for the same time at $(p - 3q)$ kilometres per hour, what distance would he cover?

7. After arranging the children in a school in $(3n - 2)$ rows with $(2n + 3)$ children in each row, $(13 - 3n)$ children remained. How many children were in the school?

8. A man earns on average a salary of $(2x - 25)$ dollars each month. What would be his total salary for $(x - 3)$ years?

-----------------------------------------------------------------------------------

# CHAPTER 10

## THE FOUR OPERATIONS

In arithmetic we learned that the **BOMDAS rule** should be used when the four operations (addition, subtraction, multiplication and division) are used together with brackets (or parenthesis) in a given problem.

---

### BOMDAS Rule

- **B**rackets: First remove brackets evaluating the expressions inside them

- **O**rdinals: Next expand any exponents

- **M**ultiplication + **D**ivision: Then multiply and divide in order from left to right

- **A**ddition + **S**ubtraction: Finally add and subtract in any order

---

If the problem specifies which operations are to be performed and in which order, then we need not use the rule. The exercises in this chapter can be done without the use of the BOMDAS rule.

### EXERCISE 10.1 --------------------------------------------------------------------------------

1. From the sum of $7a^2 - 4a + 2$ and $3a^3 + 2a^2 - 5a - 4$ subtract $8a^3 - 10a^2 - 9a + 2$.

2. What should be added to $4m^2 - 3n^2 + 7n - 4$ to get $2m^2 + m - 3$?

3. From what should $x^4 - 3x^2 - x + 9$ be subtracted to get $5x^3 - x^2 + 3x - 7$?

4. By how much is the product of $(x - 8)$ and $(x - 11)$ smaller than the product of $(2x + 3)$ and $(3x - 1)$?

5. Divide the sum of $a^2b - ab^2 - 3b^3$, $a^3 + 2ab^2 + 3a^2b$ and $2ab^2 - 5a^2b$ by $(a - b)$.

6. Divide by $1 + x^2$, the result of multiplying $3 + x + 3x^2 + x^3$ by $x + 1$.

7. By how much is the coefficient of x in the expansion of $(2x - 3)^3$ less than the coefficient of x in the expansion of $(3x - 2)^2$?

8. By how much is $(x + y + z)^2$ greater than $x^2 + y^2 + z^2 - xy + yz + zx$?

---------------------------------------------------------------------------------------------------------

## EXERCISE 10.2----------------------------------------------------------------------------

1. From 81 metres long roll of wire 9 pieces of 'a' metres and 8 pieces of '3a' metres were cut out. What is the length of the wire left over?

2. A is x years old. B is 4 years younger than A. What would be the total of their ages after 7 years?

3. Three boys weigh $2x^2 + y^2$ kg, $3x^2 - 2y^2$ kg and $x^2 + 4y^2$ kg respectively. What is their average weight? (Hint: Add the weights and divide by 3).

4. A page of a book has on average $x^2$ lines and $3y^2$ words per line. How many pages are in the book if the total number of words is $15x^3y^2$?

5. Divide \$(8ab – 3bc) between Ann and Mary so that Mary gets \$(2ab + bc) more than Ann.

6. I am 'm' years old now. My father is 'n' years older than me. What was my father's age 5 years ago? What was the total of their ages then?

7. The perimeter of a square plot of land is 12b metres. What is the area of the land?

8. What would be cost of y eggs if x dozen eggs cost \$48?

-------------------------------------------------------------------------------------------

**EXERCISE 10.3**------------------------------------------------------------------------------

1. Calculate the total cost of $(x - y)$ kilograms of tea at $\$x$ per kg and $(x + y)$ kilograms of coffee at $\$y$ per kg.

2. A rectangular field is $3a$ metres long and $2b$ metres wide. How many metres would a person cover by walking once along all four borders of the field?

3. Divide $124 among A and B so that B gets three times the amount of A.

4. The area of a rectangular plot of land is $x^2 - 4x + 3$ square metres. If the length of the land is $(x - 1)$ metres, what is its width?

5. The area of a square field is $4x^2 + 4x + 1$ square metres. What is its perimeter?

6. A man sold a piece of land for $\$(4x^2 - 3x - 5)$ and made a profit of $\$(3x - 2)$. How much did the land cost him?

7. A box is $(x^2 + 1)$ metres long, $(x - 1)$ metres wide and $(x + 1)$ metres high. What is its volume?

8. A room is $4a$ metres long, $3a$ metres wide and $6a$ metres high. Calculate the total surface area of the four walls.

--------------------------------------------------------------------------------

| Solutions: Exercise 10.3 | |
|---|---|
| **1.** $\$(x^2 + y^2)$ | **5.** $(8x + 4)$ metres |
| **2.** $(6a + 4b)$ metres | **6.** $\$(4x^2 - 6x - 3)$ |
| **3.** A: $31, B:$93 | **7.** $(x^4 - 1)m^3$ |
| **4.** $(x - 3)$ metres | **8.** $84a$ square metres |

# CHAPTER 11

## REMOVAL OF BRACKETS

Three brackets (or parenthesis) are generally used in algebra. They are:

- Simple brackets   ( )

- Curly brackets   { }

- Square brackets   [ ]

The outer most brackets are the square ones, followed by curly brackets and then finally the simple brackets.

When removing brackets the simple brackets are removed first, followed by the curly brackets and then finally square brackets as shown in the example below:

**Example 1:**

Simplify: x − [x + {x − (x + 1)}]

= x − [x + {x − x − 1}]
= x − [x − 1]
= 1

If another bracket is required within a simple bracket, this can be done by using a line as shown below. The line is to be removed first and then the brackets.

**Example 2:**

Simplify: [3x + y − {x + 4y − (2x + y − $\overline{x + 1}$)}]

= [3x + y − {x + 4y − (2x + y − x − 1)}]
= [3x + y − {x + 4y − 2x − y + x + 1}]
= [3x + y − x − 4y + 2x + y − x − 1]
= 3x − 2y − 1

Simplify by removing the brackets:

1.  $3(x - 2) + 2(x + 3)$

2.  $6(p + 4) - 5(3 + p)$

3.  $m - (2m - 5n) - 3(m + n)$

4.  $3x - (2x + y) - (x + 5y)$

5.  $5 - \{x - (y + x)\} - y$

6.  $2[p - 2\{q + (p - 2q)\}]$

7.  $(x - y) - \{x - y - (x + y) - (x - y)\}$

8.  $2m - n - [m - (m - 2n) - \{3m - (m - n)\}]$

------------------------------------------------------------------------------

**Solutions: Exercise 11.1**

| **1.** $5x$ | **2.** $p + 9$ | **3.** $-4m + 2n$ | **4.** $-6y$ | **5.** $5$ | **6.** $-2p + 4q$ | **7.** $2x$ | **8.** $4m - 2n$ |

Simplify by removing the brackets:

1.  $8x^2 - \{3xy + (2y^2 - 3x^2 - \overline{2xy + y^2})\}$

2.  $2x^2 - y(3x + y) - \{x^2 - y(3x - y)\}$

3.  $8m^2 + 2[m^2 - 2(m^2 - 3) - 6m^2]$

4.  $a^2 - (a^2 + b^2 + c^2) - (a^2 - b^2 + c^2) + (a^2 + b^2 - c^2)$

5.  $2(x^2 - y^2) - 3[x^2 - \{y^2 + xy + y(y - \overline{x - y})\}]$

6.  $x^2[y^2 - z^2 + x^2 - \{y^2 - (x^2 - \overline{y^2 - z^2} + x^2 - y^2 + z^2)\}]$

7.  $6x^2 - 5y^2 - [4x^2 - 3\{y^2 - 2(x^2 - y^2)\}]$

8.  $m^2 - [n^2 - 2\{n^2 - 3(m^2 - n^2 + 1) - 3\} - 7]$

------------------------------------------------------------------------------

**Solutions: Exercise 11.2**

| **1.** $11x^2 - y^2 - xy$ | **3.** $-6m^2 + 12$ | **5.** $-x^2 + 7y^2$ | **7.** $-4x^2 + 4y^2$ |
| **2.** $x^2 - 2y^2$ | **4.** $b^2 - 3c^2$ | **6.** $x^4 - x^2z^2$ | **8.** $-5m^2 + 7n^2 - 5$ |

When a number is multiplied by itself, the result is called the square of that number. Similarly, when the square of a number is multiplied again by the number the result is the number cubed.

---

Taking 8 to represent our number, the above statements can be written as follows:

- $8 \times 8 = 8^2$

- $8^2 \times 8 = 8 \times 8 \times 8 = 8^3$

The number 8 can be written as $8^1$, but it is customary to omit the '1'.

---

**8** is the **square root of $8^2$**.and the **cube root of $8^3$.**

---

The square root and cube root of numbers are represented by symbols as shown below:

- The symbol for square root is $\sqrt{\phantom{x}}$

- The symbol for the cube root is $\sqrt[3]{\phantom{x}}$

---

In a manner similar to finding the square root and cube root of numbers, we can also find square root and cube root of algebraic numbers. For example, the square root of $x^2$ is x. and the cube root of $x^3$ is also x.

---

**a** is the **square root of $a^2$** and the **cube root of $a^3$.**

---

The square root of the algebraic term $25x^4$ would be $5x^2$, and the cube root of the algebraic term $216x^6$ would be $6x^2$. In other words, $5x^2 \times 5x^2 = 25x^4$ and $6x^2 \times 6x^2 \times 6x^2 = 216x^6$.

Note that the square root of a number can be positive or negative.

---

**Example:**

$\sqrt{64} = 8$ or $-8$   (This is because both +8 and −8 when squared will give 64)

---

The quickest way to find the **square root** of a number is to express it first as a product of its prime factors.

**Example:**

- Find the positive square root of 64

$64 = \underline{2 \times 2} \times \underline{2 \times 2} \times \underline{2 \times 2}$

Now, determine how many **pairs** of the prime factor '2' are there. There are three pairs of '2'. Therefore:

$\sqrt{64} = 2 \times 2 \times 2 = 8$

However, if we were to find the square root of 32, we would not get a whole number square root.

If we express 32 as a product of its prime factors:

$32 = \underline{2 \times 2} \times \underline{2 \times 2} \times 2$

There are two pairs of 2's and a single 2. Therefore:

$\sqrt{32} = 2 \times 2 \times \sqrt{2} = 4\sqrt{2}$

Similarly, the positive square root of 108 can be written as follows:

$\sqrt{108} = \sqrt{2 \times 2 \times 3 \times 3 \times 3} = \sqrt{2 \times 2} \times \sqrt{3 \times 3} \times \sqrt{3} = 2 \times 3\sqrt{3} = 6\sqrt{3}$

Not all numbers will have a whole number square root. The actual square roots can be obtained using a calculator or by an arithmetical method.

It is useful to remember the positive square roots of 2 and 3:

- $\sqrt{2} = 1.414$ (corrected to three decimal places)

- $\sqrt{3} = 1.732$ (corrected to three decimal places)

For example, the square root of 32 is $4\sqrt{2}$ which when expressed as decimals is:

$4\sqrt{2} = 4 \times 1.414 = 5.656$ (approximately)

The cube roots of all numbers are not necessarily whole numbers. The calculator can be used to determine their values to the desired number of decimal places.

**Examples:**

- Find the cube root of 64

$64 = \underline{2 \times 2 \times 2} \times \underline{2 \times 2 \times 2}$.

There are two **triples** of 2. Therefore:

$\sqrt[3]{64} = 2 \times 2 = 4$

- Find the cube root of 81

$81 = \underline{3 \times 3 \times 3} \times 3$

There is one triple of 3. Therefore:

$\sqrt[3]{81} = 3 \times \sqrt[3]{3} = 3 \times 1.442 = 4.326$

**More examples:**

- $\sqrt{m^2} = m \quad (m^2 = \underline{m \times m})$

- $\sqrt{9l^6} = 3l^3 \quad (9l^6 = \underline{3 \times 3} \times \underline{l \times l} \times \underline{l \times l} \times \underline{l \times l})$

## EXERCISE 12.1-------------------------------------------------------------------------------

Find the square root of the following:

| | | | |
|---|---|---|---|
| 1. $a^4$ | 3. $4m^8$ | 5. $9x^2y^2$ | 7. $25a^2b^4c^8$ |
| 2. $x^6$ | 4. $16p^{10}$ | 6. $4m^4n^6$ | 8. $36p^4q^{10}$ |

-------------------------------------------------------------------------------

**Solutions: Exercise 12.1**

| 1. $a^2$ | 2. $x^3$ | 3. $2m^4$ | 4. $4p^5$ | 5. $3xy$ | 6. $2m^2n^3$ | 7. $5ab^2c^4$ | 8. $6p^2q^5$ |
|---|---|---|---|---|---|---|---|

## EXERCISE 12.2-------------------------------------------------------------------------------

Find the cube root of the following:

| | | | |
|---|---|---|---|
| 1. $a^6$ | 3. $8m^{12}$ | 5. $27x^3y^6$ | 7. $125a^3b^6c^9$ |
| 2. $x^9$ | 4. $64p^9$ | 6. $343m^6n^{15}$ | 8. $216p^6q^{12}$ |

-------------------------------------------------------------------------------

## EXERCISE 12.3----------------------------------------------------------------

Express in the simplest form:

1. Square root of $\dfrac{a^2x^4}{b^4y^6}$

2. Square root of $\dfrac{m^4n^8}{625n^4}$

3. Square root of $32a^6b^{10}c^{12}$

4. Square root of $81x^4y^8z^{10}$

5. Cube root of $\dfrac{8a^3y^3}{27b^3y^3}$

6. Cube root of $729(x + y)^6$

7. Cube root of $-216a^9b^{12}c^6$

8. Cube root of $\dfrac{64x^6y^9}{125x^3y^6}$

--------------------------------------------------------------------------------

42

# CHAPTER 13

Any number whose exact value can only be expressed by using the $n^{th}$ root sign $\sqrt[n]{\phantom{x}}$ is called a **surd.**

**Example:**

- Express in the simplest form the cube root of 81

$\sqrt[3]{81} = 3\sqrt[3]{3}$     $(81 = \underline{3 \times 3 \times 3} \times 3)$

Since this cannot be simplified to a whole number, this is a **surd.**

In general if the $n^{th}$ root results in a **whole number**, it is **not a surd**.

**Example:**

- Express in the simplest form the fourth root of 256

$\sqrt[4]{256} = 4$     $(256 = \underline{4 \times 4 \times 4 \times 4})$

Since this has been simplified to a whole number, this is **not a surd.**

## ADDITION AND SUBTRACTION OF SURDS

The $n^{th}$ root of a number (and multiples of it) can be added as shown in the following examples:

**Examples:**

- $\sqrt[2]{5} + \sqrt[2]{5} = 2\sqrt[2]{5}$
- $\sqrt[5]{2} + 4\sqrt[5]{2} - 2\sqrt[5]{2} = 3\sqrt[5]{2}$
- $\sqrt[n]{a} + \sqrt[n]{a} = 2\sqrt[n]{a}$
- $2\sqrt[3]{3} + \sqrt[3]{3} + 4\sqrt[3]{3} . = 7\sqrt[3]{3} .$

The nth root of a number multiplied by the nth root of another number is the nth root of the **product of the two numbers**.

Multiplication of surds can sometimes result in a rational number.

**Examples:**

- $\sqrt[n]{a} \times \sqrt[n]{b} = \sqrt[n]{ab}$
- $\sqrt[4]{8} \times \sqrt[4]{20} = \sqrt[4]{160} = 2.\sqrt[4]{10}$  (a surd)
- $\sqrt[5]{2} \times \sqrt[5]{16} = \sqrt[5]{32} = 2$  (a rational number)

The $n^{th}$ root of a number multiplied by itself results in $n^{th}$ root of the **square of that number**.

**Examples:**

- $\sqrt[n]{a} \times \sqrt[n]{a} = \sqrt[n]{a^2}$
- $\sqrt[3]{8} \times \sqrt[3]{8} = \sqrt[3]{64} = 4$
- $\sqrt[6]{16} \times \sqrt[6]{25} = \sqrt[6]{400}$

However, the square root of a number multiplied by itself gives the number itself. For example $\sqrt[2]{5}$ multiplied by itself would equal 5.

In general, $\sqrt{a} \times \sqrt{a} = a$

The product of the $n^{th}$ root of a number and the $m^{th}$ root of the same number cannot be expressed in the above manner.

# THE DISTRIBUTIVE LAW OF SURDS

$$\sqrt{x}\,(\sqrt{y} + \sqrt{z}) = \sqrt{x} \times \sqrt{y} + \sqrt{x} \times \sqrt{z} = \sqrt{xy} + \sqrt{xz}$$

**Examples:**

- $\sqrt{5}\,(\sqrt{3} + \sqrt{8}) = \sqrt{15} + \sqrt{40}$

- $\sqrt{2}\,(\sqrt{8} - \sqrt{6}) = \sqrt{16} - \sqrt{12} = 4 - 2\sqrt{3}$

- $\sqrt{a} - \sqrt{b}$ is the conjugate of $\sqrt{a} + \sqrt{b}$

- $\sqrt{a} + \sqrt{b}$ is the conjugate of $\sqrt{a} - \sqrt{b}$

When $(\sqrt{a} + \sqrt{b})$ is multiplied by $(\sqrt{a} - \sqrt{b})$, the result is $(a - b)$.

**Examples:**

- $(\sqrt{3} + \sqrt{2})(\sqrt{3} - \sqrt{2}) = 3 - 2 = 1$

- $(\sqrt{12} + \sqrt{7})(\sqrt{12} - \sqrt{7}) = 12 - 7 = 5$

## FRACTIONAL SURDS AND RATIONALISATION

A fraction with a surd in the denominator (or in both the numerator and the denominator) is called a **fractional surd**. The surd in the denominator can be made a whole number by multiplying the numerator and denominator by that surd.

A **surd in the denominator** can be a simple surd of the form $\sqrt{x}$ or of any of the following forms: $a + \sqrt{b}$, $\sqrt{a} + b$, $\sqrt{a} + \sqrt{b}$

**Examples:**

Rationalise the following surds:

- $\dfrac{2}{\sqrt{3}}$

$= \dfrac{2}{\sqrt{3}} \times \dfrac{\sqrt{3}}{\sqrt{3}}$ (multiply the numerator and denominator by $\sqrt{3}$)

$= \dfrac{2\sqrt{3}}{3}$

- $\dfrac{3\sqrt{2}}{\sqrt{5}}$

$= \dfrac{3\sqrt{2}}{\sqrt{5}} \times \dfrac{\sqrt{5}}{\sqrt{5}} = \dfrac{3\sqrt{10}}{5}$ (multiply the numerator and denominator by $\sqrt{5}$)

In order to rationalise a complex denominator we multiply the numerator and denominator by the **conjugate of the denominator**.

**Examples:**

- $\dfrac{2}{\sqrt{3}+1}$ is rationalised as follows:

$$\frac{2}{\sqrt{3}+1} = \frac{2}{\sqrt{3}+1} \times \frac{\sqrt{3}-1}{\sqrt{3}-1}$$

$$= \frac{2(\sqrt{3}-1)}{3-1} = \sqrt{3}-1$$

- $\dfrac{\sqrt{7}}{\sqrt{2}+3}$ is rationalised as follows:

$$\frac{\sqrt{7}}{\sqrt{2}+3} = \frac{\sqrt{7}}{\sqrt{2}+3} \times \frac{\sqrt{2}-3}{\sqrt{2}-3}$$

$$= \frac{\sqrt{7}(\sqrt{2}-3)}{2-9} = \frac{\sqrt{7}(\sqrt{2}-3)}{-7} = \frac{\sqrt{7}(3-\sqrt{2})}{7}$$

- $\dfrac{5\sqrt{2}}{\sqrt{8}-\sqrt{3}}$ is rationalised as follows:

$$\frac{5\sqrt{2}}{\sqrt{8}-\sqrt{3}} = \frac{5\sqrt{2}}{\sqrt{8}-\sqrt{3}} \times \frac{\sqrt{8}+3}{\sqrt{8}+3}$$

$$= \frac{5\sqrt{2}(\sqrt{8}+\sqrt{3})}{8-3} = \sqrt{2}\,(\sqrt{8}+\sqrt{3})$$

$$= \sqrt{16}+\sqrt{6} = 4+\sqrt{6}$$

**EXERCISE 13.1**----------------------------------------------------------------

Write the following as simpler surds:

1. $\sqrt{12}$     3. $\sqrt{63}$     5. $\sqrt{98}$     7. $\sqrt{512}$

2. $\sqrt{45}$     4. $\sqrt{72}$     6. $\sqrt{300}$     8. $\sqrt{648}$

-----------------------------------------------------------------------------------

## Solutions: Exercise 13.1

| | | | | | | | |
|---|---|---|---|---|---|---|---|
| **1.** $2\sqrt{3}$ | **2.** $3\sqrt{5}$ | **3.** $3\sqrt{7}$ | **4.** $6\sqrt{2}$ | **5.** $7\sqrt{2}$ | **6.** $10\sqrt{3}$ | **7.** $16\sqrt{2}$ | **8.** $18\sqrt{2}$ |

## EXERCISE 13.2----------------------------------------------------------------

Simplify the following:

1. $\sqrt{a^2 b}$

2. $\sqrt{x^3 y^5}$

3. $2\sqrt{x} + 5\sqrt{x}$

4. $a\sqrt{75} - 3a\sqrt{3}$

5. $x\sqrt[3]{24} + x\sqrt[3]{192}$

6. $15y\sqrt[5]{32} - 2y\sqrt[4]{256}$

7. $\sqrt{x}(2\sqrt{x} - \sqrt{y})$

8. $2\sqrt{x}(3\sqrt{xy} + 5\sqrt{x})$

--------------------------------------------------------------------------------

## Solutions: Exercise 13.2

| | | | |
|---|---|---|---|
| **1.** $a\sqrt{b}$ | **3.** $7\sqrt{x}$ | **5.** $6x\sqrt[3]{3}$ | **7.** $2x - \sqrt{xy}$ |
| **2.** $xy^2\sqrt{xy}$ | **4.** $2a\sqrt{3}$ | **6.** $22y$ | **8.** $6x\sqrt{y} + 10x$ |

## EXERCISE 13.3----------------------------------------------------------------

Rationalise the following:

1. $\dfrac{1}{\sqrt{2}}$

2. $\dfrac{3}{\sqrt{5}}$

3. $\dfrac{\sqrt{a}}{\sqrt{b}}$

4. $\dfrac{3\sqrt{x}}{\sqrt{3y}}$

5. $\dfrac{\sqrt{m}}{\sqrt{m} + \sqrt{n}}$

6. $\dfrac{\sqrt{a} - b}{\sqrt{a} + b}$

7. $\dfrac{x}{\sqrt{xy} - \sqrt{x}}$

8. $\dfrac{\sqrt{c}}{2\sqrt{cd} + \sqrt{c}}$

--------------------------------------------------------------------------------

## Solutions: Exercise 13.3

| | | | |
|---|---|---|---|
| **1.** $\dfrac{\sqrt{2}}{2}$ | **3.** $\dfrac{\sqrt{ab}}{b}$ | **5.** $\dfrac{m - \sqrt{mn}}{m - n}$ | **7.** $\dfrac{\sqrt{xy} + \sqrt{x}}{(y-1)}$ |
| **2.** $\dfrac{3\sqrt{5}}{5}$ | **4.** $\dfrac{\sqrt{3xy}}{y}$ | **6.** $\dfrac{a - 2b\sqrt{a} + b^2}{a - b^2}$ | **8.** $\dfrac{2\sqrt{d} - 1}{4d - 1}$ |

## INDICES

The n[th] **root** of a number x, $\sqrt[n]{x}$, can also be expressed as $\mathbf{x^{1/n}}$.

**Examples:**

- $\sqrt{x} = x^{1/2}$
- $\sqrt[3]{a} = a^{1/3}$
- $\sqrt[5]{m} = m^{1/5}$

We have already seen that a variable multiplied by itself a number of times is expressed as the power of that number, for example, $a = a^1$ and $b \times b = b^2$. The numbers that are raised as powers of the variables in the above examples are positive fractional numbers.

Powers can also be negative whole numbers or negative fractional numbers. The negative expressions can be derived from following a rule to express fractions such as the ones shown in the next page.

- $\dfrac{1}{x}$ is written as $x^{-1}$

- $\dfrac{1}{x^n}$ is written as $x^{-n}$

- $\dfrac{1}{\sqrt{x}}$ which is the same as $\dfrac{1}{x^{\frac{1}{2}}}$ is written as $x^{-1/2}$

- $\dfrac{1}{\sqrt[n]{x}}$ which is the same as $\dfrac{1}{x^{\frac{1}{n}}}$ is written as $x^{-1/n}$

The power of a variable is generally referred to as its **index** or **exponent**. The use of more than one index is referred to as indices.

## RULES OF INDICES

There are several rules that are used when using indices in expressions and equations. These are shown in the following examples:

1. The product of a variable raised to different indices (**m** and **n**) is the same as the variable raised to the sum of those indices.

$$a^m \times a^n = a^{m+n}$$

**Examples:**

- $a^5 \times a^{-2} = a^3$  • $m^{1/2} \times m^{5/2} = m^3$  • $p^{-1/3} \times p^2 = p^{5/3}$

2. When a variable raised to the power **m** is divided by the same variable raised to the power **n**, the result is the variable raised to the difference between the powers.

$$a^m \div a^n = a^{m-n}$$

3. When a variable raised to the power m is further raised to the power n, the result is the variable raised to the product of the two powers m and n.

$$(a^m)^n = a^{mn}$$

4. A variable raised to a power is the same as the inverse of that variable raised to the same power but with an opposite sign.

$a^m = 1/a^{-m}$ and $a^{-m} = 1/a^m$

5. Any variable or number raised to the power zero is equal to 1.

$a^m \div a^m = a^{m-m} = a^0 = a^m/a^m = 1$

**Examples:**

- $a^0 = 1$
- $x^0 = 1$
- $1000^0 = 1$

**EXERCISE 13.4**----------------------------------------------------

Simplify the following:

1. $9^{1/2}$

2. $8^{1/3}$

3. $81^{1/4}$

4. $25^{3/2}$

5. $27^{2/3}$

6. $64^{-1/6}$

7. $3^{-2} \times 2^3$

8. $5(4^2 \times 3^3)^0$

-------------------------------------------------------------------

### Solutions: Exercise 13.4

| 1. 3 | 2. 2 | 3. 3 | 4. 125 | 5. 9 | 6. $\frac{1}{2}$ | 7. $\frac{8}{9}$ | 8. 5 |
|------|------|------|--------|------|------|------|------|

**EXERCISE 13.5**----------------------------------------------------

Simplify the following:

1. $a^2 \times a^4$

2. $4y^3 \times 5y^5$

3. $3p^6q^3 \times 2p^2q^4$

4. $5m^3n^2 \times 7m^{-2}n^0$

5. $b^7 \div b^3$

6. $12a^5 \div 3a^2$

7. $20e^8 \div 5e^{-1}$

8. $18c^3d^4 \div -6c^{-2}d^3$

-------------------------------------------------------------------

### Solutions: Exercise 13.5

| 1. $a^6$ | 2. $20y^8$ | 3. $6p^8q^7$ | 4. $35mn^2$ | 5. $b^4$ | 6. $4a^3$ | 7. $4e^9$ | 8. $-3c^5d$ |
|----------|------------|--------------|-------------|----------|-----------|-----------|-------------|

## EXERCISE 13.6----------------------------------------------------------------------------

Simplify the following:

1. $(a^3)^4$

2. $(3a^2)^3$

3. $3(2x)^4$

4. $(3p^2q^3)^2$

5. $b^2 \times (a^3b^4)^2$

6. $(2c^2d^4)^2 \div 2cd^3$

7. $8b^2 \div (2b)^2$

8. $(x^2y^3)^4 \div (x^3y^2)^3$

----------------------------------------------------------------------------

### Solutions: Exercise 13.6

| 1. $a^{12}$ | 2. $27a^6$ | 3. $48x^4$ | 4. $9p^4q^6$ | 5. $a^6b^{10}$ | 6. $2c^3d^5$ | 7. $2$ | 8. $\dfrac{y^6}{x}$ |
|---|---|---|---|---|---|---|---|

## EXERCISE 13.7----------------------------------------------------------------------

Simplify and express the answer using positive indices:

1. $a^2(a^{-2}b^3)^2$

2. $(x^2y^{-3})^2 \div x^{-6}$

3. $(3a^2b^{-3})^2 \div ab^2$

4. $(3a^{-1/2})^4 \times b^{-2}$

5. $2(x^{-2}y^6)^{1/3}$

6. $4m^{-1/3} \times 5m^{1/4}$

7. $\dfrac{a^0}{a^2b^{-3}}$

8. $9c^2 \div (6c^{-2}d)^2$

----------------------------------------------------------------------------

### Solutions: Exercise 13.7

| 1. $\dfrac{b^6}{a^2}$ | 2. $\dfrac{x^{10}}{y^6}$ | 3. $\dfrac{9a^3}{b^8}$ | 4. $\dfrac{81}{a^2b^2}$ | 5. $\dfrac{2y^2}{x^{2/3}}$ | 6. $\dfrac{20}{m^{1/12}}$ | 7. $\dfrac{b^3}{a^2}$ | 8. $\dfrac{c^6}{4d^2}$ |
|---|---|---|---|---|---|---|---|

# CHAPTER 14

We have seen a number being expressed as another number raised to a power (or an exponent), such as $16 = 2^4$. The number 2 is called the base and it is raised to the power 4. The base can be any number and so can the power to which it is raised, as long as the answer results in the number required. For example, $16 = 4^2 = 2^4$.

> The **logarithm** (or log) of a number with respect to a base is **the power to which we have to raise the base** to obtain the number.

When the base used is 10 we refer to the logarithm as a **common logarithm**. The base can be any number and the power does not need to be a whole number.

**Examples:**

- $Log_2 16 = 4$      • $Log_4 16 = 2$      • $10^{0.3010} = 2$
- $10^{0.4771} = 3$      • $10^{0.6990} = 5$

- $Log_{10}2 = 0.3010$      • $Log_{10}3 = 0.4771$      • $Log_{10}5 = 0.6990$
- $Log_{10}20 = 1.3010$      • $Log_{10}300 = 2.4771$      • $Log_{10}5000 = 3.6990$

The common logarithms of 2, 3 and 5 are therefore 0.3010, 0.4771 and 0.6990 respectively. The common logarithms of 20, 300 and 5000 will be 1.3010, 2.4771 and 3.6990 respectively. We can observe that the decimal part of the number in each case is the same as those for 2, 3 and 5.

> The number in front of each logarithm is called the **characteristic** and is obtained by subtracting 1 from the number of digits in the number whose logarithm is being determined. The decimal part of the number is called the **mantissa.**

In these common logarithms, the negative characteristic increases by 1 for every additional zero after the decimal point.

**Examples:**

- $Log_{10}0.2 = -1 + .3010$      • $Log_{10}0.3 = -1 + .4771$
- $Log_{10}0.5 = -1 + .6990$      • $Log_{10}0.02 = -2 + .3010$
- $Log_{10}0.003 = -3 + .4771$      • $Log_{10}0.0005 = -4 + .6990$

51

Converting the logarithm of a number back to the number is referred to as finding the **antilogarithm.**

**Example:**

The logarithm of 50 is 1.6990.
The antilogarithm of 1.6990 is 50.

This can be done easily using the calculator.

When the base used is 10, we refer to the logarithm as a 'common logarithm.'

When the base is a special letter '**e**', the logarithm is called a **Natural Logarithm** (or a Napierian Logarithm). '**e**' has the value **2.71828**.

When a logarithm is written without a base, it is assumed to be a common logarithm with a base of 10. For example, $\log 100 = 2$.

## CONVERTING LOGARITHMS FROM ONE BASE TO ANOTHER

We could come across situations when the logarithm of a number to a given base needs to be converted to a logarithm of another base. The conversion is done as follows:

**Example:**

$$\text{Log}_b a = \frac{\text{Log}_{10} a}{\text{Log}_{10} b} \quad \left(\text{Note that } \text{Log}_e a = \frac{\text{Log}_{10} a}{\text{Log}_{10} e}\right)$$

If we let 'a' to take the place of 10, then we have:

$$\text{Log}_b a = \frac{\text{Log}_a a}{\text{Log}_a b}$$

$\text{Log}_b a \times \text{Log}_a b = 1$  (since $\log_a a = 1$)

## THE THREE LAWS OF LOGARITHMS

- $\text{Log}(MN) = \text{Log}M + \text{Log}N$  (eg: $\text{Log}12 = \text{Log}(3 \times 4) = \text{Log}3 + \text{Log}4$)

- $\text{Log}\left(\dfrac{M}{N}\right) = \text{Log}M - \text{Log}N$  (eg. $\text{Log}\left(\dfrac{3}{4}\right) = \text{Log}3 - \text{Log}4$)

- $\text{Log}M^k = k\text{Log}M$  (eg. $\text{Log}8^3 = 3\text{Log}8$)

**EXERCISE 14.1**--------------------------------------------------------------------------------

Find the values of the following:

1. $\log_4 64$

2. $\log_3 27$

3. $\log_7 343$

4. $\log_8 512$

5. $\log_{10} 1000$

6. $\log_4 256$

7. $\log_5 125$

8. $\log_2 1024$

| Solutions: Exercise 14.1 | | | | | | | |
|---|---|---|---|---|---|---|---|
| **1.** 3 | **2.** 3 | **3.** 3 | **4.** 3 | **5.** 3 | **6.** 4 | **7.** 3 | **8.** 10 |

**EXERCISE 14.2**--------------------------------------------------------------------------------

Express the following as a relation in logarithms:

1. $32 = 2^5$

2. $625 = 5^4$

3. $243 = 3^5$

4. $216 = 6^3$

5. $a = b^n$

6. $x = y^{1/2}$

7. $p^2 = q^m$

8. $xy = a^{1/z}$

| Solutions: Exercise 14.2 | | | |
|---|---|---|---|
| **1.** $\log_2 32 = 5$ | **3.** $\log_3 243 = 5$ | **5.** $\log_b a = n$ | **7.** $\log_q p^2 = m$ |
| **2.** $\log_5 625 = 4$ | **4.** $\log_6 216 = 3$ | **6.** $\log_y x = \dfrac{1}{2}$ | **8.** $\log_a xy = \dfrac{1}{z}$ |

**EXERCISE 14.3**--------------------------------------------------------------------------------

Express the following as a relation in indices:

1. $\log_2 16 = 4$

2. $\log_3 27 = 3$

3. $\log_{10} 1 = 0$

4. $\log_9 27 = 1.5$

5. $\log_2 x = 3$

6. $\log_{25} y = 0.5$

7. $\log_{1/3} 9 = m$

8. $\log_a 4 = 2/3$

| Solutions: Exercise 14.3 | | | |
|---|---|---|---|
| **1.** $16 = 2^4$ | **3.** $1 = 10^0$ | **5.** $x = 2^3$ | **7.** $9 = \left(\dfrac{1}{3}\right)^m$ |
| **2.** $27 = 3^3$ | **4.** $27 = 9^{1.5}$ | **6.** $y = 25^{0.5}$ | **8.** $4 = a^{\frac{2}{3}}$ |

**EXERCISE 14.4**------------------------------------------------------

Simplify the following:

1. $\dfrac{\log_3 27}{\log_5 25}$

2. $\log_{10}10^5$

3. $\log_3 3^{-2}$

4. $\dfrac{\log_{10} 1000}{\log_6 36}$

5. $\dfrac{2}{3}\log_2 8 + \dfrac{1}{2}\log_{1/3} 9$

6. $\log_2(16 \times 64)$

7. $\log_2 8 \times \log_8 2$

8. $\log_{100} 1000$

----------------------------------------------------------------------

| Solutions: Exercise 14.4 | | | | | | | |
|---|---|---|---|---|---|---|---|
| 1. $\dfrac{3}{2}$ | 2. 5 | 3. −2 | 4. 1.5 | 5. 1 | 6. 10 | 7. 1 | 8. 1.5 |

**EXERCISE 14.5**------------------------------------------------------

Find the value using calculator:

1. $\log(12 \times 19)$

2. $\log(36 \times 28^2)$

3. $\log(104 \times 73 \times 26)$

4. $\log(37.5 \times 61.3 \times 46.8)$

5. $\log \dfrac{98}{23}$

6. $\log \dfrac{212 \times 18}{56 \times 37}$

7. $\log(783 \times 212)$

8. $\log \dfrac{49 \times 522}{76^2}$

----------------------------------------------------------------------

| Solutions: Exercise 14.5 | | | |
|---|---|---|---|
| 1. 2.3579 | 3. 5.2953 | 5. 0.6295 | 7. 5.2201 |
| 2. 4.4506 | 4. 5.0317 | 6. 0.2652 | 8. 0.6462 |

**EXERCISE 14.6**----------------------------------------------------------------------------------

Find the logarithm of the following using the calculator:

1.  342

3.  86940

5.  0.0585

7.  0.0287

2.  4673

4.  87.8

6.  0.00356

8.  0.7345

| Solutions: Exercise 14.6 | |
|---|---|
| **1.** 2.5340 | **5.** $-2 + 0.7671 = -1.2329$ |
| **2.** 3.6696 | **6.** $-3 + 0.5514 = -2.4486$ |
| **3.** 4.9392 | **7.** $-2 + 0.4579 = -1.5421$ |
| **4.** 1.9435 | **8.** $-1 + 0.8660 = -0.1340$ |

**EXERCISE 14.7**----------------------------------------------------------------------------------

Find the antilogarithm of the following using the calculator:

1.  2.2568

3.  $-3 + 0.3479$

5.  $-3 + 0.6990$

7.  $4 + 0.2346$

2.  $-2 + 0.0872$

4.  $-1 + 0.5269$

6.  $-2 + 0.3028$

8.  $2 + 0.7793$

| Solutions: Exercise 14.7 | | | |
|---|---|---|---|
| **1.** 180.63 | **3.** 0.002228 | **5.** 0.005 | **7.** 17163 |
| **2.** 0.0122 | **4.** 0.3364 | **6.** 0.02008 | **8.** 601.59 |

**EXERCISE 14.8**----------------------------------------------------------------------------------

Use the change of base rule for logarithms to express the following in terms of logarithm to the base 10.  Simplify the answer:

1.  $\log_5 8$

3.  $\log_4 18$

5.  $\log_2 48$

7.  $\log_9 1000$

2.  $\log_6 12$

4.  $\log_3 80$

6.  $\log_8 100$

8.  $\log_e 16$

| Solutions: Exercise 14.8 | | | |
|---|---|---|---|
| **1.** 1.2920 | **3.** 2.0850 | **5.** 5.5850 | **7.** 3.1439 |
| **2.** 1.3869 | **4.** 3.9887 | **6.** 2.2146 | **8.** 2.7726 |

# CHAPTER 15

An algebraic expression can be simplified to a numerical value by substituting the separate numerical values of the variables contained in it.

---

**Example:**

- Calculate the value of a + 2b + 3c if a = 1, b = 2 and c = 3

a + 2b + 3c = 1 + (2 x 2) + (3 x 3) = 14

**More examples:**

If a = 2, b= –1, c = 1, calculate the values of the following:

- a + 3b – c

= 2 + (3 × –1) – 1 = – 2

- $2a^2b + 4bc + 5a^2c$

= 2 x $(2)^2$ × (–1) + 4 × (–1) × 1 + 5 × $(2)^2$ × 1

= –8 – 4 + 20 = 8

- (a + 2b)(b – 3c)

= {2 + (2 × –1)}{–1 – (3 × 1)}

= {2 – 2}{–1 – 3} = 0 x –4 = 0

If a = –3, b = 3, c = 1 and x = –1, calculate the values of:

- $\dfrac{a^2 + b^2}{x}$

= $\dfrac{(-3)^2 + 3^2}{-1}$

= $\dfrac{9+9}{-1}$ = –18

---

- $\dfrac{a^3 - b^3}{c^3 - x^3}$

$= \dfrac{(-3)^3 - (3)^3}{1^3 - (-1)^3}$

$= \dfrac{-27 - 27}{1 - (-1)}$

$= \dfrac{-54}{2}$

$= -27$

- $\sqrt{a^2 + b^2 + c^2 - 3x^2}$

$= \sqrt{(-3)^2 + 3^2 + 1^2 - 3(1)^2}$

$= \sqrt{9 + 9 + 1 - 3}$

$= \sqrt{16}$

$= 4$

## EXERCISE 15.1------------------------------------------------------------------------------

If a = 2, b = 3, c = – 1 and d = 0, evaluate the values of the following:

1. $a + b - c$

2. $2a - b + 3c + d$

3. $a^2 + b^2 - c^2 + d^2$

4. $a^2b + b^2c + c^2d$

5. $abc - a^2 - b^2 - c^2 - d^2$

6. $(a - b)(b - c)$

7. $(a + b + c)(a - b - d)$

8. $(a^2 + b^2)(c^2 + d^2)$

--------------------------------------------------------------------------------------------------

| Solutions: Exercise 15.1 | | | | | | | |
|---|---|---|---|---|---|---|---|
| **1.** 6 | **2.** –2 | **3.** 12 | **4.** 3 | **5.** –20 | **6.** –4 | **7.** –4 | **8.** 13 |

**EXERCISE 15.2**----------------------------------------------------------------------

If x = 4, y = 5 and z = 7, find the values of the following:

1. $\dfrac{x^2 + z^2}{y}$

2. $\dfrac{z^2 - y^2}{6x}$

3. $\dfrac{x + y^2 + z}{x + z}$

4. $\dfrac{3x + y^2 - z}{6y}$

5. $\dfrac{x^3 + y^3}{9z}$

6. $\dfrac{5x + 8y^2}{11xy}$

7. $\dfrac{xy + yz + zx - 13}{xyz}$

8. $\dfrac{x^2 + y^2 + z^2}{yz}$

--------------------------------------------------------------------------------

| Solutions: Exercise 15.2 | | | | | | | |
|---|---|---|---|---|---|---|---|
| **1.** 13 | **2.** 1 | **3.** $3\frac{3}{11}$ | **4.** 1 | **5.** 3 | **6.** 1 | **7.** $\frac{1}{2}$ | **8.** $2 + \frac{4}{7}$ |

**EXERCISE 15.3**----------------------------------------------------------------------

If a = 2, b = 3, c = 4 and d = −1, evaluate the values of the following:

1. $\sqrt{2ab^2c}$

2. $\sqrt{b^2 + c^2}$

3. $\sqrt{6abc}$

4. $\sqrt{2a^2 + d^2} - \sqrt{a - 2d}$

5. $\sqrt[3]{-9abcd}$

6. $\sqrt{a^3 + b^3 + c^3 - d^3}$

7. $a^3 - \sqrt[3]{4(bc - 4d)}$

8. $\sqrt{\dfrac{abc + 1}{bc + 4}}$

--------------------------------------------------------------------------------

| Solutions: Exercise 15.3 | | | | | | | |
|---|---|---|---|---|---|---|---|
| **1.** 12 | **2.** 5 | **3.** 12 | **4.** 1 | **5.** 6 | **6.** 10 | **7.** 4 | **8.** $1\frac{1}{4}$ |

# CHAPTER 16

## FACTORISATION

When an algebraic expression is a sum of variables and the variable terms have common factors, these factors are isolated from the rest as the following examples illustrate. This is called isolating the common factors.

**Examples:**

- $3x - 12y = 3(x - 4y)$

- $5c^4 + 10bc^2 = 5c^2(c^2 + 2b)$

- $7lm^3n^2 - 28lmn = 7lmn(m^2n - 4)$

- $3x^3y^2 - 6xy^2 + 9xy = 3xy(x^2y - 2y + 3)$

- $a(a + c) + b(a + c) = (a + c)(a + b)$

- $x^2(m - 3) - (3 - m) = x^2(m - 3) + (m - 3) = (m - 3)(x^2 + 1)$

Note that in order to make $(m - 3)$ common to both the terms of the given algebraic expression, $-(3 - m)$ is written as $(m - 3)$.

## EXERCISE 16.1

Factorise by isolating the common factors:

1. $2x + 6y$
2. $3a - 9b$
3. $ax - x$
4. $ab + a$
5. $x^4 + 2x^3$
6. $a^2 - 2a^2b$
7. $x^2y^2 - x^4y^4$
8. $4c^4 - 12c^2d$

| Solutions: Exercise 16.1 | | | |
|---|---|---|---|
| **1.** $2(x + 3y)$ | **3.** $x(a - 1)$ | **5.** $x^3(x + 2)$ | **7.** $x^2y^2(1 - x^2y^2)$ |
| **2.** $3(a - 3b)$ | **4.** $a(b + 1)$ | **6.** $a^2(1 - 2b)$ | **8.** $4c^2(c^2 - 3d)$ |

**EXERCISE 16.2**------------------------------------------------------------------

Factorise by isolating the common factors:

1. $6m^3n^4 + 12m^4n^3$

2. $3p^2q^2r^3 + 9pq^2r$

3. $12ab^3c^2 - 18b^2c^2$

4. $6x^3y^2z + 24xyz$

5. $21ax^2y^4 + 14ax^3y^3$

6. $13lm^3n^2 - 26lm^2n^3$

7. $9a^3b^2x^4 - 27a^2bx^3$

8. $15p^4q^3r^3 + 5p^3q^2r^4$

| Solutions: Exercise 16.2 | | | |
|---|---|---|---|
| **1.** $6m^3n^3(n + 2m)$ | **3.** $6b^2c^2(2ab - 3)$ | **5.** $7ax^2y^3(3y + 2x)$ | **7.** $9a^2bx^3(abx - 3)$ |
| **2.** $3pq^2r(pr^2 + 3)$ | **4.** $6xyz(x^2y + 4)$ | **6.** $13lm^2n^2(m - 2n)$ | **8.** $5p^3q^2r^3(3pq + r)$ |

**EXERCISE 16.3**------------------------------------------------------------------

Factorise:

1. $4x + 8y - 12$

2. $7a + 14b + 21c$

3. $2am - 6an + 4al$

4. $a^2 - a^3 + a^4$

5. $4p^3 + 8p^2 + 20p$

6. $x^3 + x^2y - xy^2$

7. $8x^2 + 20x + 4$

8. $2xy^2 - 6x^2y - 2x^2$

| Solutions: Exercise 16.3 | | | |
|---|---|---|---|
| **1.** $4(x + 2y - 3)$ | **3.** $2a(m - 3n + 2l)$ | **5.** $4p(p^2 + 2p + 5)$ | **7.** $4(2x^2 + 5x + 1)$ |
| **2.** $7(a + 2b + 3c)$ | **4.** $a^2(1 - a + a^2)$ | **6.** $x(x^2 + xy - y^2)$ | **8.** $2x(y^2 - 3xy - x)$ |

**EXERCISE 16.4**------------------------------------------------------------------

Factorise:

1. $5x^2y^2 - 10xy^2 + 15xy$

2. $2am^4 - 6a^2m^2 + 8am^2$

3. $8x^4 - 8x^2y^2 + 16x^2y^2$

4. $14a^3b^4 + 28a^4b^2 - 7a^3b^2$

5. $18l^2m^3n + 27lm^2n^2 - 9l^3m^3n^3$

6. $2x^2yz^3 - 8xy^2z^2 - 6yz^3$

7. $3x^3y + 2x^2y^2 - 4x^2yz$

8. $6a^4bc - 3a^2b^3c - 3a^2bc^3$

| Solutions: Exercise 16.4 | | | |
|---|---|---|---|
| **1.** $5xy(xy - 2y + 3)$ | **3.** $8x^2(x^2 - y^2 + 2y^2)$ | **5.** $9lm^2n(2lm + 3n - l^2mn^2)$ | **7.** $x^2y(3x + 2y - 4z)$ |
| **2.** $2am^2(m^2 - 3a + 4)$ | **4.** $7a^3b^2(2b^2 + 4a - 1)$ | **6.** $2yz^2(x^2z - 4xy - 3z)$ | **8.** $3a^2bc(2a^2 - b^2 - c^2)$ |

**EXERCISE 16.5**---------------------------------------------------------------------------------

Factorise:

1.     $(a + b)x + (a + b)y$

2.     $a(x + y) - (x + y)$

3.     $x^2(a - 2b) - 2(a - 2b)$

4.     $5(a - b) - (a - b)c$

5.     $(3x + y)a^2 + (3x + y)b^2$

6.     $x^2(a - 2) - (2 - a)$

7.     $m^2(b - c) + 3(c - b)$

8.     $p^2(l - m) - 3(m - l)$

---------------------------------------------------------------------------------

| Solutions: Exercise 16.5 | | | |
|---|---|---|---|
| **1.** $(a + b)(x + y)$ | **3.** $(a - 2b)(x^2 - 2)$ | **5.** $(3x + y)(a^2 + b^2)$ | **7.** $(b - c)(m^2 - 3)$ |
| **2.** $(x + y)(a - 1)$ | **4.** $(a - b)(5 - c)$ | **6.** $(a - 2)(x^2 + 1)$ | **8.** $(l - m)(p^2 + 3)$ |

**EXERCISE 16.6**---------------------------------------------------------------------------------

Factorise:

1.     $x^2(2a - b) + 3(b - 2a)$

2.     $m(x + a) + n(x + a) - (x + a)$

3.     $a(x - y) + b(y - x) + c(x - y)$

4.     $m(p + 3q) - n(p + 3q) + l(3q + p)$

5.     $x(a + b + c) - y(a + b + c) - a - b - c$

6.     $l(x - y - z) + m(y + z - x) - (x - y - z)$

7.     $a(p + q - r) - b(r - p - q) - c(p + q - r)$

8.     $a(x^2 + y^2 - z^2) + b(-x^2 - y^2 + z^2) + (z^2 - x^2 - y^2)$

---------------------------------------------------------------------------------

| Solutions: Exercise 16.6 | | | |
|---|---|---|---|
| **1.** $(2a - b)(x^2 - 3)$ | **3.** $(x - y)(a - b + c)$ | **5.** $(a + b + c)(x - y - 1)$ | **7.** $(p + q - r)(a + b - c)$ |
| **2.** $(x + a)(m + n - 1)$ | **4.** $(p + 3q)(m - n + l)$ | **6.** $(x - y - z)(l - m - 1)$ | **8.** $(x^2 + y^2 - z^2)(a - b - 1)$ |

## FACTORISATION BY GROUPING TERMS

Sometimes it is necessary to group terms in pairs to facilitate taking out a common factor.  The following examples illustrate this point.

**Examples:**

- Factorise $ax + bx + ay + by$

$\underline{ax + bx} + \underline{ay + by}$    (The terms grouped are underlined)

$= x(a + b) + y(a + b)$

$= (a + b)(x + y)$

- Factorise $ab + 1 - b - a$

$ab + 1 - b - a$

$= \underline{ab - a} - \underline{b + 1}$    (The terms grouped are underlined)

$= a(b - 1) - 1(b - 1)$    (Note the change in sign)

$= (b - 1)(a - 1)$

- Factorise $3xy - 4y - 6x + 8$

$\underline{3xy - 4y} - \underline{6x + 8}$

$= y(3x - 4) - 2(3x - 4)$    (Note the change in sign)

$= (3x - 4)(y - 2)$

- Factorise $m^3 - m^2 + m - 1$

$\underline{m^3 - m^2} + \underline{m - 1}$

$= m^2(m - 1) + 1(m - 1)$

$= (m - 1)(m^2 + 1)$

- Factorise $(x + 3y)^2 - 2x - 6y$

$(x + 3y)^2 - 2x - 6y$

$= (x + 3y)(x + 3y) - 2(x + 3y)$

$= (x + 3y)(x + 3y - 2)$

**EXERCISE 16.7**------------------------------------------------

Factorise by grouping terms:

| | |
|---|---|
| 1. $ax + ay + bx + by$ | 5. $2ax + 2ay - 2bx - 2by$ |
| 2. $x^2 + xy + xz + yz$ | 6. $m^2 + 2m + 2n + mn$ |
| 3. $m^2 + mn + m + n$ | 7. $p^2 - px + pq - qx$ |
| 4. $a^2 + ab - a - b$ | 8. $x^3 + x^2 - 3x - 3$ |

------------------------------------------------

**Solutions: Exercise 16.7**

| | | | |
|---|---|---|---|
| **1.** $(x + y)(a + b)$ | **3.** $(m + n)(m + 1)$ | **5.** $2(x + y)(a - b)$ | **7.** $(p - x)(p + q)$ |
| **2.** $(x + y)(x + z)$ | **4.** $(a + b)(a - 1)$ | **6.** $(m + 2)(m + n)$ | **8.** $(x + 1)(x^2 - 3)$ |

**EXERCISE 16.8**------------------------------------------------

Factorise by grouping terms:

| | |
|---|---|
| 1. $a^4 - a^3 - 2a + 2$ | 5. $2a^2b - 1 - a^2 + 2b$ |
| 2. $(x + 2)^2 - 5x - 10$ | 6. $x^2 - 1 + 3x^5 - 3x^3$ |
| 3. $x^2 - (2 - 5y)x - 10y$ | 7. $ax + bx - a^2 - ab + cx - ac$ |
| 4. $3x^2 - y(x + 1) + 3x$ | 8. $ax + bx - a^2 - ab - x + a$ |

------------------------------------------------

**Solutions: Exercise 16.8**

| | | | |
|---|---|---|---|
| **1.** $(a - 1)(a^3 - 2)$ | **3.** $(x - 2)(x + 5y)$ | **5.** $(a^2 + 1)(2b - 1)$ | **7.** $(x - a)(a + b + c)$ |
| **2.** $(x + 2)(x - 3)$ | **4.** $(3x - y)(x + 1)$ | **6.** $(x^2 - 1)(1 + 3x^3)$ | **8.** $(x - a)(a + b - 1)$ |

## FACTORISATION OF QUADRATIC EXPRESSIONS

A **quadratic expression** is an algebraic expression with terms containing powers of the variable, the highest power being 2.

**Examples:**

- Factorise $2x^2 + 11x + 5$

- The middle term of this quadratic expression is $11x$. This is really $11x^1$.
- The number 5 is actually $5x^0$, $x^0$ has a value of 1 and so $5x^0 = 5$
- The powers of x involved in this expression are thus: 2, 1 and 0

Observe that in the quadratic expression $2x^2 + 11x + 5$ the coefficient of $x^2$ is 2, the coefficient of x is 11 and the constant is 5. To factorise this quadratic expression, we do the following steps:

1.  Multiply the coefficient of $x^2$, 2, by the constant to give 10.

2.  Split 10 into two factors so that their product is 10 and their sum is 11. The answer is 10 and 1.

3.  Rewrite the quadratic expression by replacing 11x with 10x and 1x as follows: $\underline{2x^2 + 10x + 1x + 5}$

4.  Factorise the underlined grouped terms and write the results as follows: $2x(x + 5) + 1(x + 5)$

5:  Decide that the factors are $(x + 5)$ and $(2x + 1)$.
    Therefore $2x^2 + 11x + 5 = (x + 5)(2x + 1 )$

- Factorise $y^2 – 7y + 12$

1.  Multiply the coefficient of $y^2$, 1, by the constant 12 to give 12.

2.  Split 12 into two factors so that their product is 12 and their sum is –7 The answer is –4 and –3.

3.  Rewrite the quadratic expression by replacing –7y with –4y and –3y as follows: $\underline{y^2 – 4y – 3y + 12}$

4.  Factorise the underlined grouped terms and write the results as follows: $y(y – 4) – 3(y – 4)$

5.  Decide that the factors are $(y – 4)$ and $(y – 3)$.
    Therefore $y^2 – 7y + 12 = (y – 4)(y – 3)$

- Factorise $6a^2 + 13a – 28$

1.  Multiply the coefficient of $a^2$, 6, by the constant –28 to give –168.

2.  Split –168 into two factors so that their product is –168 and their sum is 13. The answer is –8 and 21.

3.  Rewrite the quadratic expression by replacing 13a with –8a and 21a as follows: $\underline{6a^2 – 8a + 21a – 28}$

4.  Factorise the underlined grouped terms and write the results as follows: $2a(3a – 4) + 7(3a – 4)$

5.  Decide that the factors are $(3a – 4)$ and $(2a + 7)$.
    Therefore $6a^2 + 13a – 28 = (3a – 4)(2a + 7)$

**EXERCISE 16.9**--------------------------------------------------------------------------------

Factorise the following quadratic expressions:

1. $a^2 + 3a + 2$
3. $m^2 + 5m + 6$
5. $p^2 + 4p + 4$
7. $x^2 + 12x + 35$

2. $x^2 + 5x + 4$
4. $a^2 + 10a + 9$
6. $y^2 + 6y + 9$
8. $p^2 + 8p + 7$

--------------------------------------------------------------------------------

| Solutions: Exercise 16.9 | | | |
|---|---|---|---|
| **1.** $(a + 1)(a + 2)$ | **3.** $(m + 2)(m + 3)$ | **5.** $(p + 2)(p + 2)$ | **7.** $(x + 5)(x + 7)$ |
| **2.** $(x + 1)(x + 4)$ | **4.** $(a + 1)(a + 9)$ | **6.** $(y + 3)(y + 3)$ | **8.** $(p + 1)(p + 7)$ |

**EXERCISE 16.10**-----------------------------------------------------------------------------

Factorise the following quadratic expressions by inspection:

1. $y^2 + 4y + 3$
3. $r^2 + 10r + 9$
5. $m^2 + 19m + 48$
7. $x^2 + 12x + 35$

2. $m^2 + 9m + 18$
4. $b^2 + 8b + 12$
6. $p^2 + 21p + 68$
8. $x^2 + 32x + 87$

--------------------------------------------------------------------------------

| Solutions: Exercise 16.10 | | | |
|---|---|---|---|
| **1.** $(y + 1)(y + 3)$ | **3.** $(r + 1)(r + 9)$ | **5.** $(m + 3)(m + 16)$ | **7.** $(x + 5)(x + 7)$ |
| **2.** $(m + 3)(m + 6)$ | **4.** $(b + 2)(b + 6)$ | **6.** $(m + 4)(m + 17)$ | **8.** $(x + 3)(x + 29)$ |

**EXERCISE 16.11**-----------------------------------------------------------------------------

Factorise the following quadratic expressions:

1. $x^2 - 4x + 3$
3. $p^2 - 14p + 40$
5. $y^2 - 10y - 56$
7. $c^2 - 13c - 30$

2. $x^2 - 6x + 8$
4. $m^2 + m - 12$
6. $b^2 - 2b - 15$
8. $x^2 - 7x - 30$

--------------------------------------------------------------------------------

| Solutions: Exercise 16.11 | | | |
|---|---|---|---|
| **1.** $(x - 1)(x - 3)$ | **3.** $(p - 4)(p - 10)$ | **5.** $(y + 4)(y - 14)$ | **7.** $(c + 2)(c - 15)$ |
| **2.** $(x - 2)(x - 4)$ | **4.** $(m + 4)(m - 3)$ | **6.** $(b + 3)(b - 5)$ | **8.** $(x + 3)(x - 10)$ |

**EXERCISE 16.12**-----------------------------------------------------------------------------

Factorise the following quadratic equations by inspection:

1. $3 - 2a - a^2$
3. $p^2 + p - 20$
5. $z^2 + 3z - 10$
7. $y^2 - 17y + 42$

2. $2 - x - x^2$
4. $a^2 - 10a + 21$
6. $k^2 - 16k + 64$
8. $54 - 15x - x^2$

--------------------------------------------------------------------------------

**EXERCISE 16.13**-------------------------------------------------------------------

Factorise the following quadratic expressions:

1. $2x^2 + 3x + 1$     3. $2m^2 + 7m + 3$     5. $3p^2 + 5p + 2$     7. $4x^2 + 11x - 3$

2. $2x^2 + 9x + 4$     4. $2a^2 + 5a + 2$     6. $3x^2 - 11x + 6$     8. $3a^2 - 5a - 2$

----------------------------------------------------------------------------------

| Solutions: Exercise 16.13 | | | |
|---|---|---|---|
| **1.** $(2x + 1)(x + 1)$ | **3.** $(2m + 1)(m + 3)$ | **5.** $(3p + 2)(p + 1)$ | **7.** $(4x - 1)(x + 3)$ |
| **2.** $(2x + 1)(x + 4)$ | **4.** $(2a + 1)(a + 2)$ | **6.** $(3x - 2)(x - 3)$ | **8.** $(3a + 1)(a - 2)$ |

**EXERCISE 16.14**-------------------------------------------------------------------

Factorise the following quadratic expressions by inspection:

1. $2 - 5x + 3x^2$     3. $2x^2 + 3x - 2$     5. $6a^2 - 13a + 6$     7. $3 - 2y - 21y^2$

2. $5 - 2m - 7m^2$     4. $6p^2 - 11p + 3$     6. $9b^2 - 5b - 14$     8. $7x^2 + 47x - 14$

----------------------------------------------------------------------------------

| Solutions: Exercise 16.14 | | | |
|---|---|---|---|
| **1.** $(2 - 3x)(1 - x)$ | **3.** $(2x - 1)(x + 2)$ | **5.** $(2a - 3)((3a - 2)$ | **7.** $(3 + 7y)(1 - 3y)$ |
| **2.** $(5 - 7m)(1 + m)$ | **4.** $(3p - 1)(2p - 3)$ | **6.** $(9b - 14)(b + 1)$ | **8.** $(7x - 2)(x + 7)$ |

**EXERCISE 16.15**-------------------------------------------------------------------

Factorise the following quadratic expressions:

1.     $x^2 - 3xy + 2y^2$          5.     $2a^2 + 13ab + 15b^2$

2.     $a^2 + 4ab - 21b^2$          6.     $6c^2 - 7cd - 5d^2$

3.     $x^2 + 3xy - 18y^2$          7.     $10p^2 - 13pq - 3q^2$

4.     $m^2 - 6mn + 8n^2$          8.     $21a^2 + 4ab - 32b^2$

----------------------------------------------------------------------------------

| Solutions: Exercise 16.15 | | | |
|---|---|---|---|
| **1.** $(x - 2y)(x - y)$ | **3.** $(x + 6y)(x - 3y)$ | **5.** $(2a + 3b)(a + 5b)$ | **7.** $(5p + q)(2p - 3q)$ |
| **2.** $(a + 7b)(a - 3b)$ | **4.** $(m - 2n)(m - 4n)$ | **6.** $(3c - 5d)(2c + d)$ | **8.** $(3a + 4b)(7a - 8b)$ |

**EXERCISE 16.16**----------------------------------------------------------------------------------------

Factorise the following quadratic expressions:

1. $4a^2 - ab - 14b^2$

2. $5x^2 - 38xy + 21y^2$

3. $2x^2 - 19xy + 17y^2$

4. $12m^2 - 20mn + 3n^2$

5. $6a^2 - 2ab - 28b^2$

6. $56k^2 - 19kp - 10p^2$

7. $22a^2 + 75ab - 7b^2$

8. $3x^2 - 25xy + 52y^2$

----------------------------------------------------------------------------------------

| Solutions: Exercise 16.16 | | | |
|---|---|---|---|
| **1.** $(a - 2b)(4a + 7b)$ | **3.** $(2x - 17y)(x - y)$ | **5.** $(a + 2b)(6a - 14b)$ | **7.** $(2a + 7b)(11a - b)$ |
| **2.** $(x - 7y)(5x - 3y)$ | **4.** $(2m - 3n)(6m - n)$ | **6.** $(7k + 2p)(8k - 5p)$ | **8.** $(x - 4y)(3x - 13y)$ |

## Difference of two squares

The difference of the squares of two numbers is equal to the product of their sum and their difference.

**Example 1:** $x^2 - y^2 = (x + y)(x - y)$

In some expressions that could be factorised using the difference of two squares method, a common factor needs to be taken out first.

**Example 2:** $3a^2 - 3b^2 = 3(a^2 - b^2) = 3(a + b)(a - b)$

In other expressions one or both terms in the expression may have to be put in a complete square form before factorising using the method of difference of two squares.

**More examples:**

- $16p^2 - q^2 = (4p)^2 - q^2 = (4p + q)(4p - q)$

- $25x^2y^2 - 9z^2 = (5xy)^2 - (3z)^2 = (5xy + 3z)(5xy - 3z)$

- $4(a + b)^2 - c^2 = (2a + 2b)^2 - c^2$

$= (2a + 2b + c)(2a + 2b - c)$

- $9a^2 - 16(b + c)^2 = (3a)^2 - (4b + 4c)^2$

$= (3a + 4b + 4c)(3a - 4b - 4c)$

**EXERCISE 16.17**------------------------------------------------------------------------

Factorise using difference of two squares:

1. $m^2 - n^2$       3. $a^2 - 49$       5. $x^2 - 121$       7. $9x^2 - 1$

2. $p^2 - q^2$       4. $1 - x^2$       6. $x^2y^2 - 9$       8. $16 - x^2$

------------------------------------------------------------------------

| Solutions: Exercise 16.17 | | | |
|---|---|---|---|
| **1.** $(m + n)(m - n)$ | **3.** $(a + 7)(a - 7)$ | **5.** $(x + 11)(x - 11)$ | **7.** $(3x + 1)(3x - 1)$ |
| **2.** $(p + q)(p - q)$ | **4.** $(1 + x)(1 - x)$ | **6.** $(xy + 3)(xy - 3)$ | **8.** $(4 + x)(4 - x)$ |

**EXERCISE 16.18**------------------------------------------------------------------------

Factorise using difference of two squares:

1. $16x^2 - 4y^2$       3. $a^4 - 9b^4$       5. $36x^2y^2 - 25z^2$       7. $x^4y^6 - z^2$

2. $49x^2y^2 - 25$       4. $4 - x^4y^2$       6. $25 - 4m^6$       8. $16 - p^8q^2$

------------------------------------------------------------------------

| Solutions: Exercise 16.18 | | | |
|---|---|---|---|
| **1.** $4(2x + y)(2x - y)$ | **3.** $(a^2 + 3b^2)(a^2 - 3b^2)$ | **5.** $(6xy + 5z)(6xy - 5z)$ | **7.** $(x^2y^3 + z)(x^2y^3 - z)$ |
| **2.** $(7xy + 5)(7xy - 5)$ | **4.** $(2 + x^2y)(2 - x^2y)$ | **6.** $(5 + 2m^3)(5 - 2m^3)$ | **8.** $(4 + p^4q)(4 - p^4q)$ |

**EXERCISE 16.19**------------------------------------------------------------------------

Factorise using difference of two squares:

1. $2x^2 - 50$       3. $x^3 - 16x$       5. $27a^2 - 12a^4$       7. $4x^8 - x^6$

2. $2a^2 - 8b^2$       4. $49a^2x^2 - a^2$       6. $2x^3 - 18xy^2$       8. $52x^2y^2z - 13a^2b^2z$

------------------------------------------------------------------------

| Solutions: Exercise 16.19 | | | |
|---|---|---|---|
| **1.** $2(x + 5)(x - 5)$ | **3.** $x(x + 4)(x - 4)$ | **5.** $3a^2(3 + 2a)(3 - 2a)$ | **7.** $x^6(2x + 1)(2x - 1)$ |
| **2.** $2(a + 2b)(a - 2b)$ | **4.** $a^2(7x + 1)(7x - 1)$ | **6.** $2x(x + 3y)(x - 3y)$ | **8.** $13z(2xy + ab)(2xy - ab)$ |

**EXERCISE 16.20**------------------------------------------------------------------------

Factorise using difference of two squares:

1. $(x + y)^2 - z^2$       3. $(m - n)^2 - 4$       5. $16x^4 - (x + 1)^2$       7. $a^2 - 4(x + y)^2$

2. $(b + c)^2 - d^2$       4. $x^4 - (y + z)^2$       6. $(a + b)^2 - (x - y)^2$       8. $4b^2 - (b^2 + c^2)^2$

------------------------------------------------------------------------

**EXERCISE 16.21**------------------------------------------------------------------------

Factorise using difference of two squares:

1.  $9 - 4(x + y)^2$

2.  $1 - 16(m - n)^2$

3.  $(3x - 2y)^2 - 4(x - y)^2$

4.  $a^2 - b^2 - 4(a - b)$

5.  $x^2 - 4y^2 - x + 2y$

6.  $4m^2(m + 1) - (m + 1)$

7.  $4x^2 - y^2 - 2xy - y^2$

8.  $4x^2 + 4xy + y^2 - 1$

----------------------------------------------------------------------------

**EXERCISE 16.22**------------------------------------------------------------------------

Simplify using the difference of two squares:

1.  $(28)^2 - (22)^2$

2.  $(56\frac{1}{2})^2 - (33\frac{1}{2})^2$

3.  $(101)^2 - 1$

4.  $(337)^2 - (163)^2$

5.  $(73)^2 - (27)^2$

6.  $(84\frac{3}{4})^2 - (15\frac{1}{4})^2$

7.  $(163.52)^2 - (36.48)^2$

8.  $(178.43)^2 - (78.43)^2$

----------------------------------------------------------------------------

# CHAPTER 17

## HIGHEST COMMON FACTOR

In a manner similar to finding the highest common factor (HCF) of two or more numbers, we can find the highest common factor of two or more algebraic expressions.

The **highest common factor** is obtained by multiplying together all the factors common to the given numbers.

**Example:**

- Find the highest common factor of 36 and 54.

$36 = 2 × 2 × 3 × 3$
$54 = 2 × 3 × 3 × 3$
$HCF = 2 × 3 × 3 = 18$   (Note that the factor 3 is common twice)

**More examples:**

- Find the highest common factor of $6x^3y^2$ and $18x^4y$

$6x^3y^2 = 2 × 3 × x × x × x × x × y × y$
$18x^4y = 2 × 3 × 3 × x × x × x × x × x × y$
$HCF = 2 × 3 × x × x × x × x × y = 6x^3y$

- Find the highest common factor of $x^2 - y^2$ and $x^2 - xy$

$x^2 - y^2 = (x + y) × (x - y)$
$x^2 - xy = x × (x - y)$
$HCF = (x - y)$

- Find the highest common factor of $x^3 + y^3$ and $x^3 - x^2y + xy^2$

$x^3 + y^3 = (x + y) × (x^2 - xy + y^2)$
$x^3 - x^2y + xy^2 = x × (x^2 - xy + y^2)$
$HCF = (x^2 - xy + y^2)$

- Find the highest common factor of $x^2 + 3x +3$, $x^2 - x - 6$ and $2x^2 + 3x - 2$

$x^2 + 3x +2 = (x + 2) × (x + 1)$
$x^2 - x - 6 = (x + 2) × (x - 3)$
$2x^2 + 3x - 2 = (x + 2) × (2x - 1)$
$HCF = x + 2$

**EXERCISE 17.1**--------------------------------------------------------------------------

Find the highest common factor of the following:

1.    16, 24

2.    27, 63

3.    32, 48

4.    8x, 12x

5.    12x, 15x

6.    20ab, 35bc

7.    18lm, 27mn

8.    63xy, 35xz

--------------------------------------------------------------------------

**Solutions: Exercise 17.1**

| **1.** 8 | **2.** 9 | **3.** 16 | **4.** 4x | **5.** 3x | **6.** 5b | **7.** 9m | **8.** 7x |
|---|---|---|---|---|---|---|---|

**EXERCISE 17.2**--------------------------------------------------------------------------

Find the HCF of the following:

1.    $7a^3b^2, 21a^2b^3$

2.    $6p^2q^3, 9pq^4$

3.    $9x^2y, 12y^2x$

4.    $30x^3y^2z^3, 42xy^4z^4$

5.    $12a^4b^2c^3, 18a^3bc^2$

6.    $18p^5q^4r^3, 27p^6q^3r^2$

7.    $8am^2x, 6a^2mx, 4amx^2$

8.    $5x^3y^2z^2, 15x^2y^3z^2, 20x^2y^2z^3$

--------------------------------------------------------------------------

**Solutions: Exercise 17.2**

| **1.** $7a^2b^2$ | **3.** $3xy$ | **5.** $6a^3bc^2$ | **7.** $2amx$ |
|---|---|---|---|
| **2.** $3pq^3$ | **4.** $6xy^2z^3$ | **6.** $9p^5q^3r^2$ | **8.** $5x^2y^2z^2$ |

**EXERCISE 17.3**--------------------------------------------------------------------------

Find the HCF of the following:

1.    $x(x + y), x(x - y)$

2.    $2a + 2b, 3a + 3b$

3.    $3x - 6y, 4x - 8y$

4.    $a^2 - b^2, a^2 + ab$

5.    $a^2 - b^2, a^2 - ab$

6.    $(p + q)^2, p^2 - q^2$

7.    $x^3 - y^3, x^2 - y^2$

8.    $ax + x, a^4x + a^3x$

--------------------------------------------------------------------------

## Solutions: Exercise 17.3

| **1.** x | **3.** x − 2y | **5.** a − b | **7.** (x − y) |
|---|---|---|---|
| **2.** a + b | **4.** a + b | **6.** p + q | **8.** x(a + 1) |

## EXERCISE 17.4------------------------------------------------------------------------------

Find the HCF of the following:

1.  $2x^2 + x - 3, 2x^2 + 5x + 3$

2.  $x^3 - 36x, x^3 + 2x^2 - 48x$

3.  $3m^2 + 7m - 6, 2m^2 + 7m + 3$

4.  $a^2 + 3ab, a^3 - 9ab^2, a^3 + 6a^2b + 9ab^2$

5.  $x^3 + 1, x^2 - 1, x^2 - 2x - 3$

6.  $4m^2 - 1, 4m^2 + 8m + 3, 6m^2 + m - 1$

7.  $x^2 - x - 6, x^2 - 6x + 9, x^2 + x - 12$

8.  $9x^2 - 64, 6x^2 + 5x - 56, 3x^2 - 23x + 40$

------------------------------------------------------------------------------

## Solutions: Exercise 17.4

| **1.** 2x + 3 | **3.** m + 3 | **5.** x + 1 | **7.** x − 3 |
|---|---|---|---|
| **2.** x(x − 6) | **4.** a(a + 3b) | **6.** 2m + 1 | **8.** 3x − 8 |

# CHAPTER 18

## LOWEST COMMON MULTIPLE

In a manner similar to finding the lowest common multiple (LCM) of two or more numbers, we can find the lowest common multiple of two or more algebraic expressions.

> The **lowest common multiple** of a set of numbers is the smallest number that can be divided exactly by all the individual members of the set.

**Example:**

- Find the lowest common multiple of 28 and 42.

$28 = 2 \times 2 \times 7$
$42 = 2 \times 3 \times 7$

The lowest common multiple is obtained by multiplying together all the factors common to all given numbers and all remaining factors of all the numbers. In this example, the common factors are 2 and 7 and the remaining factors are 2 and 3.

Therefore, LCM = $2 \times 7 \times 2 \times 3 = 84$

**More examples:**

Find the lowest common multiple of:

- $3a^3b$, $9a^2b^2$ and $12a^3b^3$

$3a^3b = 3 \times a \times a \times a \times b$
$9a^2b^2 = 3 \times 3 \times a \times a \times b \times b$
$12a^3b^3 = 2 \times 2 \times 3 \times a \times a \times a \times b \times b \times b$

Here, we isolate the factors common to all three expressions, then any factor common to two expressions and multiply them together with any remaining factors.

LCM $= 3 \times a \times a \times b \times b \times 3 \times 2 \times 2 \times a \times b = 36a^3b^3$

- $x^2 + 2xy$ and $x^2 - 4y^2$

$x^2 + 2xy = x \times (x + 2y)$
$x^2 - 4y^2 = (x + 2y) \times (x - 2y)$

$LCM = (x + 2y) \times x \times (x - 2y)$
$= x(x^2 - 4y^2)$

- $a^2 - 1$, $a^2 + 3a + 2$, $a^2 + a - 2$

$a^2 - 1 = (a + 1)(a - 1)$
$a^2 + 3a + 2 = (a + 2)(a + 1)$
$a^2 + a - 2 = (a + 2)(a - 1)$

$LCM = (a + 2)\,(a + 1)\,(a - 1)$

## EXERCISE 18.1------------------------------------------------------------------

Find the lowest common multiple of the following:

1.  12, 32          3.   35, 63          5.   $6x$, $8x$          7.   $xy$, $xy^2$

2.  18, 42          4.   $5x$, $3x$          6.   $9a$, $15a$          8.   $2x^2$, $5xy$

------------------------------------------------------------------

| Solutions: Exercise 18.1 | | | | | | | |
|---|---|---|---|---|---|---|---|
| **1.** 96 | **2.** 126 | **3.** 315 | **4.** $15x$ | **5.** $24x$ | **6.** $45a$ | **7.** $xy^2$ | **8.** $10x^2y$ |

## EXERCISE 18.2------------------------------------------------------------------

Find the lowest common multiple of the following:

1.  $4xy$, $3yz$

2.  $27x^2y^2$, $36x^3y^3$

3.  $pqrs$, $mnpq$

4.  $4x^2y$, $3xz$, $6yz^2$

5.  $a^2p^3q$, $a^3pq^2$, $bp^2q^2$

6.  $6x^3y$, $9x^2y^3$, $12x^3y^3$

7.  $15a^4b^5$, $18a^2b^3$, $6a^3b^4$

8.  $14m^6n^3$, $7m^4n^2$, $28m^5n^4$

------------------------------------------------------------------

| Solutions: Exercise 18.2 | | | |
|---|---|---|---|
| **1.** $12xyz$ | **3.** $mnpqrs$ | **5.** $a^3bp^3q^2$ | **7.** $90a^4b^5$ |
| **2.** $108x^3y^3$ | **4.** $12x^2yz^2$ | **6.** $36x^3y^3$ | **8.** $28m^6n^4$ |

**EXERCISE 18.3**----------------------------------------------------------------

Find the lowest common multiple of the following:

1. $2x + 10, 3x + 15$

2. $a^2 - ab, ab - ac$

3. $xy^2 + xy, y^2 + y$

4. $2x^3 - 6x^2y, 3x^2 - 9xy$

5. $m^2 - 2mn, m^2 - 4n^2$

6. $1 - 2a + a^2, a - a^2$

7. $1 - 2x + x^2, 1 - x^2$

8. $a^2 - ab, (a - b), (a - c)$

----------------------------------------------------------------

| Solutions: Exercise 18.3 | | | |
|---|---|---|---|
| **1.** $6(x + 5)$ | **3.** $xy(y + 1)$ | **5.** $m(m + 2n)(m - 2n)$ | **7.** $(1 - x)^2(1 + x)$ |
| **2.** $a(a - b)(b - c)$ | **4.** $6x^2(x - 3y)$ | **6.** $a(1 - a)^2$ | **8.** $a(a - b)(a - c)$ |

**EXERCISE 18.4**----------------------------------------------------------------

Find the lowest common multiple of the following:

1. $xy^2, (a + b)^2, x^2y(a + b)^3$

2. $2m^2 + 4mn, m^2 - 4n^2$

3. $x^2 + 3x + 2, x^2 + x - 2$

4. $(x - 1)^2, x(x + 1), x^2 + 2x + 1$

5. $2x^2 - 5x + 3, 6x^2 - 5x - 6, 3x^2 - x - 2$

6. $x^2 - 1, x^2 + 3x + 2, x^2 + x - 2$

7. $2x^2 + x - 1, x^2 + 3x + 2, 2x^2 + 3x - 2$

8. $x^3 + 2x^2 + x + 2, x^2 + 3x + 2, x^2 + x - 2$

----------------------------------------------------------------

| Solutions: Exercise 18.4 | |
|---|---|
| **1.** $x^2y^2(a + b)^3$ | **5.** $(2x - 3)(3x + 2)(x - 1)$ |
| **2.** $2m(m + 2n)(m - 2n)$ | **6.** $(x + 1)(x - 1)(x + 2)$ |
| **3.** $(x + 2)(x + 1)(x - 1)$ | **7.** $(2x - 1)(x + 1)(x + 2)$ |
| **4.** $x(x + 1)^2(x - 1)^2$ | **8.** $(x^2 + 1)(x + 2)(x + 1)(x - 1)$ |

# CHAPTER 19

## ADDITION AND SUBTRACTION

The addition and subtraction of algebraic fractions are done by finding the LCM of the denominators or first expressing the various fractions as equivalent fractions with the same denominator, similar to what we do in Arithmetic.

**Example:**

Evaluate: $\dfrac{2}{a} + \dfrac{3}{b}$

LCM of a and b = ab

Divide 'a' into 'ab' and multiply the result by 2 to get 2b
Divide 'b' into 'ab' and multiply the result by 3 to get 3a

Therefore $\dfrac{2}{a} + \dfrac{3}{b} = \dfrac{2b + 3a}{ab}$

If we were to express $\dfrac{2}{a}$ and $\dfrac{3}{b}$ as equivalent fractions with the same denominator, they would be $\dfrac{2b}{ab}$ and $\dfrac{3a}{ab}$. We can now simply add the numerators together and the result is divided by the common denominator 'ab' to get the answer $\dfrac{2b + 3a}{ab}$.

**More examples:**

- Evaluate : $\dfrac{1}{x^2} - \dfrac{1}{xy} + \dfrac{1}{y^2}$

LCM = $x^2 y^2$

Therefore, $\dfrac{1}{x^2} - \dfrac{1}{xy} + \dfrac{1}{y^2} = \dfrac{y^2 - xy + x^2}{x^2 y^2}$

- Evaluate: $\dfrac{x-2}{3} + \dfrac{x+5}{5}$

LCM = 15

Therefore, $\dfrac{x-2}{3} + \dfrac{x+5}{5} = \dfrac{5(x-2)+3(x+5)}{15}$

$= \dfrac{5x-10+3x+15}{15} = \dfrac{8x+5}{15}$

- Evaluate: $\dfrac{a}{a-b} - 1$

LCM = a – b

Therefore, $\dfrac{a}{a-b} - 1 = \dfrac{a-(a-b)}{a-b}$

$= \dfrac{a-a+b}{a-b} = \dfrac{b}{a-b}$

- Evaluate: $\dfrac{1}{x+2} - \dfrac{1}{x-4}$

LCM = (x + 2)(x – 4)

Therefore, $\dfrac{1}{x+2} - \dfrac{1}{x-4} = \dfrac{(x-4)-(x+2)}{(x+2)(x-4)}$

$= \dfrac{x-4-x-2}{(x+2)(x-4)} = \dfrac{-6}{(x+2)(x-4)}$

- Evaluate: $\dfrac{4}{x-y} + \dfrac{3}{y-x}$

LCM = x – y

$\dfrac{4}{x-y} + \dfrac{3}{y-x} = \dfrac{4}{x-y} - \dfrac{3}{x-y} = \dfrac{1}{x-y}$

- Evaluate: $\dfrac{1}{2x-3y} - \dfrac{x+y}{4x^2-9y^2}$

Factorising the denominator of the second fraction we have:

$$\dfrac{1}{2x-3y} - \dfrac{x+y}{4x^2-9y^2} = \dfrac{1}{2x-3y} - \dfrac{x+y}{(2x+3y)(2x-3y)}$$

LCM = $(2x + 3y)(2x - 3y)$

$$\dfrac{1}{2x-3y} - \dfrac{x+y}{(2x+3y)(2x-3y)} = \dfrac{(2x+3y)-(x+y)}{(2x+3y)(2x-3y)}$$

$$= \dfrac{2x+3y-x-y}{(2x+3y)(2x-3y)} = \dfrac{x+2y}{(2x+3y)(2x-3y)}$$

**EXERCISE 19.1**--------------------------------------------------------------

Evaluate the following

1. $\dfrac{1}{x} + \dfrac{1}{y}$

2. $\dfrac{2}{a} + \dfrac{3}{b}$

3. $1 + \dfrac{2}{x}$

4. $3 + \dfrac{2}{m}$

5. $\dfrac{1}{a} - \dfrac{1}{b}$

6. $1 - \dfrac{2}{a}$

7. $\dfrac{x}{y} - 1$

8. $\dfrac{1}{2x} - 1$

----------------------------------------------------------------------

| **Solutions: Exercise 19.1** | | | |
|---|---|---|---|
| 1. $\dfrac{x+y}{xy}$ | 3. $\dfrac{x+2}{x}$ | 5. $\dfrac{b-a}{ab}$ | 7. $\dfrac{x-y}{y}$ |
| 2. $\dfrac{3a+2b}{ab}$ | 4. $\dfrac{3m+2}{m}$ | 6. $\dfrac{a-2}{a}$ | 8. $\dfrac{1-2x}{2x}$ |

**EXERCISE 19.2**--------------------------------------------------------------

Evaluate the following:

1. $\dfrac{1}{a} + \dfrac{1}{b} + \dfrac{1}{c}$

2. $\dfrac{2}{x} - \dfrac{1}{y} - \dfrac{3}{z}$

3. $\dfrac{1}{2x} + \dfrac{1}{3y}$

4. $\dfrac{a}{b} - \dfrac{b}{a}$

5. $\dfrac{1}{3a} - \dfrac{1}{b} + \dfrac{1}{2c}$

6. $\dfrac{1}{x^2} + \dfrac{1}{xy}$

7. $\dfrac{a}{xy} - \dfrac{b}{yz}$

8. $\dfrac{1}{a^2} - \dfrac{1}{ab} + \dfrac{1}{b^2}$

----------------------------------------------------------------------

| Solutions: Exercise 19.2 | | | |
|---|---|---|---|
| 1. $\dfrac{ab+bc+ac}{abc}$ | 3. $\dfrac{3x+2y}{6xy}$ | 5. $\dfrac{2bc-6ac+3ab}{6abc}$ | 7. $\dfrac{az-bx}{xyz}$ |
| 2. $\dfrac{2yz-xz-3xy}{xyz}$ | 4. $\dfrac{a^2-b^2}{ab}$ | 6. $\dfrac{x+y}{x^2y}$ | 8. $\dfrac{b^2-ab+a^2}{a^2b^2}$ |

## EXERCISE 19.3

Simplify the following:

1. $\dfrac{x+1}{3} + \dfrac{x+2}{3}$

5. $\dfrac{3}{2} - \dfrac{a+3}{2}$

2. $\dfrac{2x+1}{2} + \dfrac{x+1}{3}$

6. $\dfrac{7a-2}{2} + \dfrac{2a+3}{4} - 4$

3. $\dfrac{a-2}{3} + \dfrac{a-5}{5}$

7. $\dfrac{4x-2}{3} - 4 - \dfrac{x-1}{3}$

4. $\dfrac{3x+1}{6} + \dfrac{4x-1}{4}$

8. $\dfrac{3a-2}{4} - \dfrac{a+1}{2} - 2$

| Solutions: Exercise 19.3 | | | |
|---|---|---|---|
| 1. $\dfrac{2x+3}{3}$ | 3. $\dfrac{8a-25}{15}$ | 5. $\dfrac{-a}{2}$ | 7. $\dfrac{3x-13}{3}$ |
| 2. $\dfrac{8x+5}{6}$ | 4. $\dfrac{18x-1}{12}$ | 6. $\dfrac{16a-17}{4}$ | 8. $\dfrac{a-12}{4}$ |

## EXERCISE 19.4

Simplify the following:

1. $\dfrac{2}{a} + \dfrac{3}{a+1}$

5. $\dfrac{1}{a} - \dfrac{1}{a+2b}$

2. $\dfrac{a}{a+b} + 1$

6. $\dfrac{1}{a-3} + \dfrac{1}{a-2}$

3. $\dfrac{x}{x-y} - 1$

7. $1 - \dfrac{a^2}{a^2+b^2}$

4. $\dfrac{1}{x} + \dfrac{1}{x+3y}$

8. $\dfrac{x^2}{x^2-y^2} - 1$

| Solutions: Exercise 19.4 | | | | | | |
|---|---|---|---|---|---|---|
| **1.** $\dfrac{5a+2}{a(a+1)}$ | | **3.** $\dfrac{y}{x-y}$ | | **5.** $\dfrac{2b}{a(a+2b)}$ | | **7.** $\dfrac{b^2}{a^2+b^2}$ |
| **2.** $\dfrac{2a+b}{a+b}$ | | **4.** $\dfrac{2x+3y}{x(x+3y)}$ | | **6.** $\dfrac{2a-5}{(a-3)(a-2)}$ | | **8.** $\dfrac{y^2}{x^2-y^2}$ |

## EXERCISE 19.5

Simplify the following:

1. $\dfrac{a+x}{a-x} + \dfrac{a-x}{a+x}$

5. $\dfrac{1}{a+b} + \dfrac{1}{a-b} + \dfrac{1}{a^2+ab}$

2. $\dfrac{1}{a-1} + \dfrac{1}{a+1} + \dfrac{2}{a^2-1}$

6. $\dfrac{1}{2x^2-x-3} + \dfrac{1}{2x^2+x-1}$

3. $\dfrac{1}{m-2} + \dfrac{1}{m^2-4} - \dfrac{1}{(m-2)^2}$

7. $\dfrac{1}{2x-3y} - \dfrac{x+y}{4x^2-9y^2}$

4. $\dfrac{1}{x-y} - \dfrac{1}{x+y} - \dfrac{2y}{x^2-y^2}$

8. $\dfrac{3x+1}{x^2+x-2} - \dfrac{2x-3}{x^2-3x+2}$

| Solutions: Exercise 19.5 | | |
|---|---|---|
| **1.** $\dfrac{2(a^2+x^2)}{(a+x)(a-x)}$ | | **5.** $\dfrac{2a^2+a-b}{a(a^2-b^2)}$ |
| **2.** $\dfrac{2}{a-1}$ | | **6.** $\dfrac{4(x-1)}{(2x-3)(x+1)(2x-1)}$ |
| **3.** $\dfrac{m^2-8}{(m^2-4)(m-2)}$ | | **7.** $\dfrac{x+2y}{4x^2-9y^2}$ |
| **4.** 0 | | **8.** $\dfrac{x^2-6x+4}{(x-2)(x+2)(x-1)}$ |

## MULTIPLICATION, DIVISION AND SIMPLIFICATION

In a manner similar to that we do with numbers, multiplication and division are also done in algebra but we may need to factorise expressions before we can proceed to simplify.

**Examples:**

- Simplify: $\dfrac{12xy}{3x}$

$$\dfrac{12xy}{3x} = \dfrac{4 \times 3 \times x \times y}{3 \times x}$$

= 4y   (3 and x in the denominator cancel with 3 and x in the numerator)

- Simplify: $\dfrac{x^2 + x}{x^2 - 1}$

We first factorise both the numerator and denominator:

$$\dfrac{x^2 + x}{x^2 - 1} = \dfrac{x(x+1)}{(x+1)(x-1)}$$

$= \dfrac{x}{x-1}$   (x + 1 in the numerator and denominator cancel out)

- Simplify: $\dfrac{3x^2}{x^2 - y^2} \times \dfrac{x + y}{x}$

We factorise $x^2 - y^2$ first to get $(x + y)(x - y)$.

Therefore, $\dfrac{3x^2}{x^2 - y^2} \times \dfrac{x + y}{x} = \dfrac{3x^2}{(x+y)(x-y)} \times \dfrac{x+y}{x}$

$= \dfrac{3x}{x-y}$   (Cancelling out x + y and x):

- Simplify: $\dfrac{x^2 - x - 2}{x^2 - 5x + 6} \div \dfrac{x^2 + 3x + 2}{x^2 - 3x}$

$$\dfrac{x^2 - x - 2}{x^2 - 5x + 6} \div \dfrac{x^2 + 3x + 2}{x^2 - 3x} = \dfrac{(x+1)(x-2)}{(x-2)(x-3)} \div \dfrac{(x+1)(x+2)}{x(x-3)}$$

$$= \dfrac{(x+1)(x-2)}{(x-2)(x-3)} \times \dfrac{x(x-3)}{(x+1)(x+2)} = \dfrac{x}{x+2}$$

## EXERCISE 19.6 --------------------------------------------------------------------------------

Express the following fractions in their simplest form:

1. $\dfrac{9x^2}{3x}$

2. $\dfrac{x^2 y^2}{x^3 y}$

3. $\dfrac{12x^3 y^2 z^4}{18x^3 y^3 z^2}$

4. $\dfrac{x^2 + xy}{x^3 - xy}$

5. $\dfrac{3a^2 - 4a}{3ab^2 - 4b^2}$

6. $\dfrac{x^2 - y^2}{x^3 - y^3}$

7. $\dfrac{a^2 - 1}{3a^2 + 3a}$

8. $\dfrac{xy}{x^2 y^2 + xy}$

| Solutions: Exercise 19.6 | | | |
|---|---|---|---|
| **1.** $3x$ | **3.** $\dfrac{2z^2}{3y}$ | **5.** $\dfrac{a}{b^2}$ | **7.** $\dfrac{a-1}{3a}$ |
| **2.** $\dfrac{y}{x}$ | **4.** $\dfrac{x+y}{x^2-y}$ | **6.** $\dfrac{x+y}{x^2+xy+y^2}$ | **8.** $\dfrac{1}{xy+1}$ |

## EXERCISE 19.7 --------------------------------------------------------------------------------

Express the following fractions in their simplest form:

1. $\dfrac{x^3}{x^2 - 4} \times \dfrac{x - 2}{x^2}$

2. $\dfrac{a}{a^2 - 9} \times \dfrac{a + 3}{a}$

3. $\dfrac{x^2 - y^2}{xy} \times \dfrac{x}{x + y}$

4. $\dfrac{m^2 - 9}{2m + 2} \times \dfrac{2}{m + 3}$

5. $\dfrac{a^2 - b^2}{ab} \div \dfrac{a + b}{a}$

6. $\dfrac{a - b}{a^2 + ab} \div \dfrac{a^2 - ab}{a^2 - b^2}$

7. $\dfrac{2x^2 - 3x - 2}{x^2 - x - 6} \times \dfrac{3x^2 - 8x - 3}{3x^2 - 5x - 2}$

8. $\dfrac{x^2 + 3x + 2}{x^2 - 4x + 4} \times \dfrac{x^2 + x - 6}{x^2 - 2x - 3} \div \dfrac{x^2 + 5x + 6}{x^2 - 5x + 6}$

| Solutions: Exercise 19.7 | | | |
|---|---|---|---|
| **1.** $\dfrac{x}{x+2}$ | **3.** $\dfrac{x-y}{y}$ | **5.** $\dfrac{a-b}{b}$ | **7.** $\dfrac{2x+1}{x+2}$ |
| **2.** $\dfrac{1}{a-3}$ | **4.** $\dfrac{m-3}{m+1}$ | **6.** $\dfrac{a-b}{a^2}$ | **8.** $1$ |

82

## EXERCISE 19.8----------------------------------------------------------------------------------

Simplify the following:

1. $\dfrac{3x^3yz}{7ab^2z} \times \dfrac{14a^3b^2z^2}{6a^2xy}$

2. $\dfrac{2a}{a^2-a-2} - \dfrac{1}{a+1}$

3. $\dfrac{1}{x-1} - \dfrac{1}{1-x} + 1$

4. $\left(\dfrac{a}{b} - \dfrac{b}{a}\right) \div \left(\dfrac{b}{a} - \dfrac{a}{b}\right)$

5. $\left(\dfrac{a}{b} - \dfrac{b}{a}\right) \div \left(\dfrac{a}{b} + \dfrac{b}{a} - 2\right)$

6. $\left(1+\dfrac{1}{a-1}\right) \div \left(1-\dfrac{1}{a+1}\right)$

7. $\left(\dfrac{1}{x} + \dfrac{1}{y}\right)\left(\dfrac{1}{x+y} + \dfrac{1}{x-y}\right)$

8. $\dfrac{x^2-y^2}{x^2-2xy+y^2} \times \dfrac{x^2+xy-2y^2}{3(x+y)} \div \dfrac{2x+4y}{3x}$

----------------------------------------------------------------------------------

| Solutions: Exercise 19.8 | | | |
|---|---|---|---|
| **1.** $x^2z^2$ | **3.** $\dfrac{x+1}{x-1}$ | **5.** $\dfrac{a+b}{a-b}$ | **7.** $\dfrac{2}{y(x-y)}$ |
| **2.** $\dfrac{a+2}{(a-2)(a+1)}$ | **4.** $\dfrac{a-b}{b-a}$ | **6.** $\dfrac{a+1}{a-1}$ | **8.** $\dfrac{x}{2}$ |

> A statement connecting a variable and some number(s) is called a **simple equation.**

For example, the variable x can be connected to two numbers 3 and 10 and be written as x + 3 = 10. We interpret this as meaning that the variable x when added to 3 gives a result of 10. The sign "=" represents the word 'gives', it is called an 'equal' sign.

The statement can thus be also read as: "x plus 3 equals 10." Now we have to figure out the value of x so that when it is added to 3 it equals 10.

Since 7 + 3 makes 10, the value of x is 7.

Also because the equation involves only x, (and not any higher powers of x), we refer to it as simple equation. The equation when represented graphically will result in a line. Hence the equation is commonly called a **simple linear equation.**

An equation is like a balance. To maintain equilibrium, when some change is made to one side of the balance, the same change must be made to the other side of the balance.

A formal method of solving for the value of the variable in an equation by making changes is shown by way of the following examples:

**Examples:**

- x + 3 = 10

x + 3 − 3 = 10 − 3    (Subtract 3 from both sides)

Therefore, x = 7.

- 3x + 5 = −7

3x + 5 − 5 = −7 − 5    (Subtract 5 from both sides)
3x  = −12

$$\frac{3x}{3} = -\frac{12}{3} \quad \text{(Divide both sides by 3)}$$

x = −4

- $8x - 24 = 6x - 16$

$8x - 24 + 24 = 6x - 16 + 24$   (Add 24 to both sides)
$8x = 6x + 8$

$8x - 6x = 6x - 6x + 8$   (Subtract 6x from both sides)

$$\frac{2x}{2} = -\frac{8}{2}$$   (Divide both sides by 2)

$x = 4$

- $21 + 7(6 - x) = 5(x - 3)$

$21 + 42 - 7x = 5x - 15$   (Remove the brackets)
$63 - 7x = 5x - 15$

$63 - 63 - 7x = 5x - 15 - 63$   (Subtract 63 from both sides)
$-7x = 5x - 78$

$-7x - 5x = 5x - 5x - 78$   (Subtract 5x from both sides)
$-12x = -78$

$$\frac{-12x}{-12} = \frac{-78}{-12}$$   (Divide both sides by –12)

$x = 6.5$

**More examples:**

- The result of adding 12 to a number is the same as subtracting 12 from seven times that number.  What is the number?

Let the number be x.

Therefore, $x + 12 = 7x - 12$

We can also write this as $7x - 12 = x + 12$

$7x - 12 - x = x - x + 12$
$6x - 12 = 12$
$6x - 12 + 12 = 12 + 12$

$6x = 24$
$x = 4$

The number is 4.

- You bought 27 stamps from a post office for $10.20. If some were 10 cents stamps and others were 60 cents stamps, how many 60 cents stamps did you buy?

Let the number of 60 cents stamps be x
Therefore, the number of ten cents stamps = 27 − x

Since the total cost of all the stamps is $10.20:

$[(27 − x) × 10] + [x × 60] = 1020$
$270 − 10x + 60x = 1020$
$270 + 50x = 1020$
$270 + 50x − 270 = 1020 − 270$
$50x = 750$

$x = 15$

The number of 60 cents stamps is 15

- The length of a room is three times its width. If the perimeter of the room is 72 metres, calculate the length and width of the room.

Let the width of the room be x metres
Therefore, the length of the room = 3x metres

Perimeter = 2(length + width)

$= 2(3x + x)$
$= 8x$

$8x = 72$     (Since the perimeter is 72 metres)
$x = 9$

The length of the room is 27 metres.   (Since the length is 3x)

**EXERCISE 20.1**------------------------------------------------------------------------------

Solve the following simple equations:

1. $x + 3 = 6$

2. $x − 4 = 5$

3. $2x + 7 = 13$

4. $2x + 8 = −2$

5. $8x − 8 = 12x$

6. $5x − 15 = 4x − 8$

7. $7x = −2x + 27$

8. $2x − 7 + 4x − 11 = 0$

--------------------------------------------------------------------------------

**Solutions: Exercise 20.1**

| 1. x = 3 | 2. x = 9 | 3. x = 3 | 4. x = −5 | 5. x = −2 | 6. x = 7 | 7. x = 3 | 8. x = 3 |
|----------|----------|----------|-----------|-----------|----------|----------|----------|

**EXERCISE 20.2**------------------------------------------------------------------------------------------

Solve the following simple equations:

1.    $7x - 12 = 13 + 5x - 5$

2.    $2x = 19 - 10x + 5$

3.    $8y - 2 - 7y - 6 = 5y$

4.    $5 - 4a + 12 = a - 2a + 2$

5.    $35 - 42m = 55 - 44m$

6.    $3 - 7p - 5 = -7 - 4p$

7.    $5x - 1 = 3(2x - 3)$

8.    $4(2 - y) = 5(3 - y)$

**Solutions: Exercise 20.2**

| **1.** $x = 10$ | **2.** $x = 2$ | **3.** $y = -2$ | **4.** $a = 5$ | **5.** $m = 10$ | **6.** $p = 1\frac{2}{3}$ | **7.** $x = 8$ | **8.** $y = 7$ |
|---|---|---|---|---|---|---|---|

**EXERCISE 20.3**------------------------------------------------------------------------------------------

Solve the following simple equations:

1.  $1 + 5(6 - m) = 3(5 + m)$

2.  $3(2p - 5) - 4(p - 3) = 3(p + 1)$

3.  $10(t + 9) = 5(t + 3) - 4(t - 3)$

4.  $2q - 11 - 10(5 - q) = 2(5q - 1)$

5.  $3(x - 7) + 10x = 44 + 6(2 - x) - 3x$

6.  $2(n + 2) + 2(3 - n) - 8(n - 3) = 5(5 - n)$

7.  $m(m + 1) = m^2 + 2$

8.  $x(x - 5) = x^2 - 20$

**Solutions: Exercise 20.3**

| **1.** $m = 2$ | **2.** $p = -6$ | **3.** $t = -7$ | **4.** $q = 29\frac{1}{2}$ | **5.** $x = 3\frac{1}{2}$ | **6.** $n = 3$ | **7.** $m = 2$ | **8.** $x = 4$ |
|---|---|---|---|---|---|---|---|

**EXERCISE 20.4**------------------------------------------------------------------------------------------

Solve the following equations:

1.  $y(y + 7) - 3 = y(y - 2) + 6$

2.  $-q(1 + q) + q(3 + q) = 4$

3.  $(x + 1)(x + 3) + 3 = x^2 + 8x - 6$

4.  $(2x + 1)(3x - 1) = x(6x + 4) + 2$

5.  $(p + 2)(p + 3) - 2p(p + 1) = -(p - 2)(p - 3)$

6.  $(x + 1)^2 + 3 = (x + 5)(x + 3) - 5$

7.  $(m - 3)^2 = (m + 1)(m - 5) + 10$

8.  $(2x + 1)(x + 3) - 2(x + 1)(x - 1) = 9x + 17$

**Solutions: Exercise 20.4**

| **1.** $y = 1$ | **2.** $q = 2$ | **3.** $x = 3$ | **4.** $x = -1$ | **5.** $p = 6$ | **6.** $x = -1$ | **7.** $m = 2$ | **8.** $x = -6$ |
|---|---|---|---|---|---|---|---|

# CHAPTER 21

## SIMPLE EQUATIONS WITH FRACTIONS

When a simple equation involves fractions it is necessary to multiply the equation by the LCM of the denominators. This is done, so that the results can be expressed as a simple equation without fractions. Once converted, it can be solved in the manner studied in Chapter 20. The following examples illustrate the process:

**Examples:**

- Solve: $\dfrac{x}{9} = \dfrac{1}{3}$

The LCM of 9 and 3 is 9. Multiplying both sides of the equation by 9 we get:

$$\dfrac{x}{9} \times 9 = \dfrac{1}{3} \times 9$$

Therefore, x = 3

Another method of solving this type of equations is to cross multiply.

That is, 3 × x = 9 × 1

3x = 9

x = 3

- Solve: $4\dfrac{1}{3}y = 9\dfrac{3}{4}$

Here, first we have to express the mixed fractions as improper fractions.

That is, $\dfrac{13}{3}y = \dfrac{39}{4}$

The LCM of 3 and 4 is 12. Multiplying both sides of the equation by 12 we get:

52y = 117

$$y = \dfrac{117}{52} = \dfrac{9}{4} = 2\dfrac{1}{4}$$

- Solve: $\dfrac{6x+1}{7} + 2 = \dfrac{2x+7}{3}$

The LCM of 7 and 3 is 21.

Multiplying both sides of the equation by 21 we get:

$3(6x + 1) + 2 \times 21 = 7(2x + 7)$

$18x + 3 + 42 = 14x + 49$

$18x + 45 = 14x + 49$

$18x - 14x + 45 = 14x - 14x + 49$

$4x + 45 = 49$

$4x + 45 - 45 = 49 - 45$

$4x = 4$

$x = 1$

- Solve: $\dfrac{1}{3}x - \dfrac{1}{3}(2x + 1) = 1\dfrac{1}{3} + \dfrac{1}{5}(x - 3)$

LCM = 15

Multiplying both sides of the equation by 15 we get,

$5x - 5(2x + 1) = 20 + 3(x - 3)$

$5x - 10x - 5 = 20 + 3x - 9$

$-5x - 5 = 11 + 3x$

$-5x - 5 - 3x = 11 + 3x - 3x$

$-8x - 5 = 11$

$-8x - 5 + 5 = 11 + 5$

$-8x = 16$

$x = -2$

- Solve:  $0.5(x - 8) + 3 = 0.6(x + 5)$

The decimal fractions could be converted to common fractions first and then continue as in the previous examples.  If we work with decimal fractions we expand the brackets and continue:

$0.5x - 4 + 3 = 0.6x + 3$

$0.5x - 1 = 0.6x + 3$

$0.5x - 0.6x - 1 = 0.6x + 3 - 0.6x$

$-0.1x - 1 = 3$

$-0.1x - 1 + 1 = 3 + 1$

$0.1x = -4$

$x = -40$

## EXERCISE 21.1--------------------------------------------------------------------------------------

Solve the following equations:

1.  $\dfrac{x}{6} = \dfrac{1}{3}$

2.  $\dfrac{2x}{5} = 4$

3.  $\dfrac{1}{3}x = \dfrac{3}{4}$

4.  $\dfrac{1}{4}x - 2\dfrac{1}{2} = 0$

5.  $\dfrac{2}{7}x + 3 = 1$

6.  $\dfrac{x}{3} + \dfrac{x}{2} = 5$

7.  $\dfrac{1}{4}a + \dfrac{1}{5}a = 9$

8.  $\dfrac{m}{6} + \dfrac{m}{4} = \dfrac{m}{3} + 1$

--------------------------------------------------------------------------------------------------------

| Solutions: Exercise 21.1 | | | | | | | |
|---|---|---|---|---|---|---|---|
| **1.** $x = 2$ | **2.** $x = 10$ | **3.** $x = 2\dfrac{1}{4}$ | **4.** $x = 10$ | **5.** $x = -7$ | **6.** $x = 6$ | **7.** $a = 20$ | **8.** $m = 12$ |

**EXERCISE 21.2**------------------------------------------------------------------

Solve the following equations:

1. $1\frac{1}{5}x - 7 = 2x - 2\frac{3}{5}$

2. $\frac{p}{4} + \frac{2}{5} = \frac{3}{4} + \frac{p}{5}$

3. $\frac{6}{7} + \frac{m}{3} = \frac{m}{7} - \frac{2}{3}$

4. $\frac{k}{5} - 3\frac{1}{5} = 4\frac{1}{2} - \frac{k}{2}$

5. $y + \frac{10}{27} = \frac{1}{9}y + \frac{5}{2}$

6. $0.7(x - 5) - 0.3(x + 2) = 3$

7. $0.5(2x + 4) + 2 = 0.25x$

8. $0.8(m - 5) + 2 = 0.9(m + 7)$

---

**Solutions: Exercise 21.2**

| **1.** $x = -5\frac{1}{2}$ | **2.** $p = 7$ | **3.** $m = -8$ | **4.** $k = 11$ | **5.** $y = 2\frac{19}{48}$ | **6.** $x = 17.75$ | **7.** $x = -5.33$ | **8.** $m = -83$ |
|---|---|---|---|---|---|---|---|

**EXERCISE 21.3**------------------------------------------------------------------

Solve the following equations:

1. $\frac{5 + 3x}{6} - \frac{2 - 6x}{3} = 1$

2. $\frac{x + 5}{4} = \frac{x + 2}{6} - \frac{x + 1}{8}$

3. $\frac{3}{5}(m - 4) + 2 = \frac{1}{12}(11m - 77)$

4. $\frac{11}{12}x - \frac{1}{6}(12 + x) + 6\frac{1}{2} = 0$

5. $\frac{1}{3}(a + 2) - \frac{1}{9}(3 - a) - \frac{1}{6}(2a - 5) = 2\frac{1}{6}$

6. $\frac{9}{10}x + \frac{3}{4}(x - 2) = 3 + \frac{1}{5}(5x - 3)$

7. $\frac{1}{7}(2x - 3) + \frac{1}{3}(x + 3) = 2 - \frac{x}{3}$

8. $\frac{3}{4}(a - 1) - \frac{4}{5}(a - 6) = \frac{1}{3}(a - 4) + \frac{1}{60}$

---

## EXERCISE 21.4

1.  One fifth of a number when added to one seventh of it gives 48. What is that number?

2.  A number minus seven ninth of it gives 18. What is the number?

3.  The result of subtracting five eighth of a number from three times the number is the same as increasing the number by 22. What is the number?

4.  The sum of half, one third, one sixth and one seventh of a number is 72. What is the number?

5.  The sum of two numbers is 63. The smallest number is x. If half of the smaller number is equal to one fifth of the other number, what are the two numbers?

6.  Divide 25 into two parts such that four times the bigger part is ten less than seven times the smaller part.

7.  John's age is $\frac{11}{13}$th of that of Peter. If the sum of the ages is 48, what are their respective ages?

8.  A table costs $x. The table and a chair costs $96. The cost of the table is $28 more than eight-ninths of that of the chair. How much does the table cost?

1.   Of two consecutive numbers, three eighth of the first number is three less than three seventh of the second number. What are the two numbers?

2.   In a mixed school the number of girls is 9/14th of the number of boys. If the there are 920 children in the school, how many are boys? How many are girls?

3.   My son is 21 years younger than me at present. In nine years he will be five seventh of my age. What is my present age?

4.   A trader bought articles for $19 each and sold them all for a profit of $400. If he sold a dozen for $252, how many articles did he buy?

5.   When a boy travelled from A to B at 10 km per hour he was 30 minutes late. If he travelled at 11 km per hour he would have been only 15 minutes late. What is the distance between A and B?

6.   Divide $210 among X, Y and Z so that X gets $21 more than Y and Y gets five eighths of that of Z's share.

7.   Two machines A and B cost $2880. If machine A costs 25% more than B, what is the cost of machine A?

8.   A fruit seller sold some fruit at $38 per 100. If he had sold them at $38 per 90 he would have got $19 more. How many fruits did he have?

---

| Solutions: Exercise 21.5 | |
|---|---|
| **1.** 48, 49 | **5.** $27\frac{1}{2}$ km |
| **2.** 560 boys, 360 girls | **6.** x = $73$\frac{1}{2}$, y = $52\frac{1}{2}$, z = 84 |
| **3.** $64\frac{1}{2}$ years | **7.** $1600 |
| **4.** 200 | **8.** 450 |

# CHAPTER 22

## SURDS, MODULUS, EXPONENTS AND LOGS

## SURD EQUATIONS

We usually encounter equations with one to three surds. When there is only one surd, isolate it to one side. If there are two surds, keep one on either side of the equal sign. When there are three surds keep one on one side and the other two on the other side of the equal sign.

In all cases we have to square both sides of the equation and square as many times as needed to get rid of the surd. When we get a solution for the variable, check to see if it satisfies the equation.

**Examples:**

- Solve: $\sqrt{3x-2} + 3 = 7$

By isolating the surd to one side we get,

$\sqrt{3x-2} = 4$   (Subtracting 3 from both sides)

Squaring both sides of the equation we get:

$3x - 2 = 16$

$3x = 18$

$x = 6$

- Solve: $\sqrt{x+8} - \sqrt{x} = 2$

Rewriting the equation by keeping one surd on either side we get,

$\sqrt{x+8} = \sqrt{x} + 2$

$x + 8 = x + 4\sqrt{x} + 4$   (Squaring both sides)

$4\sqrt{x} = 4$

$\sqrt{x} = 1$

$x = 1$

- Solve $\sqrt{4x+13} - \sqrt{x} = \sqrt{x+7}$

Rewriting keeping one surd on one side and the other two on the other side we get,

$\sqrt{4x+13} = \sqrt{x} + \sqrt{x+7}$

$4x + 13 = x + x + 7 + 2\sqrt{x} \times \sqrt{x+7}$

$2x + 6 = 2\sqrt{x} \times \sqrt{x+7}$

$x + 3 = \sqrt{x} \times \sqrt{x+7}$

$x^2 + 6x + 9 = x(x + 7)$    (Squaring both sides)

$x^2 + 6x + 9 = x^2 + 7x$

$x = 9$

- Solve: $\sqrt{8m+9} = m + 2$

Squaring both sides of the equation we get:

$8m + 9 = m^2 + 4m + 4$

$m^2 - 4m - 5 = 0$

$(m - 5)(m + 1) = 0$

The values of m satisfying this equation are either 5 or – 1

Therefore, m = 5 or –1

Now we check these values to see whether they satisfy the original equation.

For m = 5;

$\sqrt{40+9} = 5 + 2$

$7 = 7$    (5 is a solution)

For m = –1;

$\sqrt{-8+9} = -1 + 2$

$1 = 1$    (–1 is a solution)

Find the solutions of the following equations

| | |
|---|---|
| 1.  $\sqrt{2x+4} = 3$ | 5.  $\sqrt{3m+4} - 2 = \sqrt{2m-4}$ |
| 2.  $\sqrt{5x-1} + 2 = 5$ | 6.  $\sqrt{3x+7} - 1 = \sqrt{2x+3}$ |
| 3.  $\sqrt{3x-7} = \sqrt{x}$ | 7.  $\sqrt{4l+2} - \sqrt{4l-2} = 2$ |
| 4.  $\sqrt{3a+1} - 3 = \sqrt{a-4}$ | 8.  $\sqrt{8y+25} = y + 2$ |

----------------------------------------------------------------------------------

| **Solutions: Exercise 22.1** | | | | | | | |
|---|---|---|---|---|---|---|---|
| **1.** 2.5 | **2.** 2 | **3.** 3.5 | **4.** 8 or 5 | **5.** 4 or 20 | **6.** 3 or -1 | **7.** $\frac{1}{2}$ | **8.** 7 |

Find the solutions of the following equations

| | |
|---|---|
| 1.  $\sqrt{x+9} = x - 3$ | 5.  $(y^2 + 3y + 3)^{1/2} = 2y + 5$ |
| 2.  $3\sqrt{x+1} = x + 3$ | 6.  $\sqrt{x+2} - \sqrt{x-1} = \sqrt{\dfrac{x}{2}}$ |
| 3.  $(a^2 - 3)^{1/2} = 3a - 5$ | 7.  $\sqrt{x+4} - 2\sqrt{x-1} = -\sqrt{2x-9}$ |
| 4.  $\sqrt{p+2} - \sqrt{p-3} = 1$ | 8.  $-\sqrt{x+3} + 2\sqrt{x+8} = \sqrt{5x+11}$ |

----------------------------------------------------------------------------------

| **Solutions: Exercise 22.2** | | | | | | | |
|---|---|---|---|---|---|---|---|
| **1.** 7 | **2.** 0 or 3 | **3.** 2 | **4.** 7 | **5.** –2 | **6.** 2 | **7.** 5 | **8.** 1 |

## MODULUS EQUATIONS

The modulus of a number is its **absolute value**.  That is, its distance from zero on a number line ignoring the sign.  The symbol used to indicate this is two vertical bars.

**Examples:**

- $|3| = 3$
- $|-x| = x$

- $|-7| = 7$
- $|-(x+2)| = x + 2$

When solving equations involving modulus, it is necessary to remove the vertical bars and put a ± sign on one side of the equation as shown below. If surds are involved, square both sides and then there is no need to put a ± sign.

**Examples:**

- Solve: $|x - 2| = |2x - 5|$

Removing the vertical bars and putting ± on one side, we get:

$x - 2 = \pm (2x - 5)$

This means, $x - 2 = (2x - 5)$ or $x - 2 = -(2x - 5)$

Therefore $x = 3$ or $x = 2.333$    (both values of x satisfy the original equation)

- Solve: $\dfrac{|2x - 1|}{\sqrt{x + 1}} = 1$

Squaring both sides and cross multiplying we get:

$(2x - 1)^2 = (x + 1)$

$x^2 - 4x + 1 = x + 1$

$x^2 - 5x = 0$

$x(x - 5) = 0$

Therefore $x = 0$ or $x = 5$

**EXERCISE 22.3**----------------------------------------------------------------------

Find the solutions (if any) of the following

1. $|6x| = 18$

2. $|a - 4| = 3$

3. $|3x + 1| = |x + 5|$

4. $|2a + 3| = |a + 6|$

5. $|y - 8| + |y| = 6$

6. $|n + 2| + |n - 2| = 4$

7. $|x| + |x - 6| = 6$

8. $|y - 8| - |y| = 6$

| Solutions: Exercise 22.3 | | | |
|---|---|---|---|
| **1.** 3 or − 3 | **3.** 2 or − 1.5 | **5.** No solution | **7.** 0 or 6 |
| **2.** 7 or 1 | **4.** 3 or − 3 | **6.** Value of n between -2 & +2 | **8.** 1 |

**EXERCISE 22.4**-------------------------------------------------------------------------------------------------------

Find the solutions (if any) of the following:

1.    $|4 - 3b| = 19$

5.    $|p - 2| = \frac{1}{2}|p - 6|$

2.    $|c - 3| = |4c + 6|$

6.    $|y - 4| = -3 + |y|$

3.    $\dfrac{|3x - 2|}{\sqrt{x + 1}} = 3.5$

7.    $\dfrac{\sqrt{x + 7}}{|x - 1|} = 3$

4.    $\dfrac{\sqrt{2x + 1}}{|x - 3|} = 3$

8.    $\dfrac{|x - 3|}{\sqrt{x + 5}} = \dfrac{1}{3}$

-------------------------------------------------------------------------------------------------------

| Solutions: Exercise 22.4 | | | |
|---|---|---|---|
| **1.** $-5$ or $7\frac{2}{3}$ | **3.** $3$ or $\frac{-11}{36}$ | **5.** $-2$ or $3\frac{1}{3}$ | **7.** $2$ or $\frac{1}{9}$ |
| **2.** $-3$ or $-\frac{3}{5}$ | **4.** $4$ or $2\frac{2}{9}$ | **6.** $3\frac{1}{2}$ | **8.** $4$ or $2\frac{1}{9}$ |

## EXPONENTIAL EQUATIONS

**Exponential functions** can be products, quotients or sums.

With products and quotients of exponential functions, we must keep all the exponentials on the left side and constants on the right. If it is possible to express the exponentials and constants in the same base we should do that and solve the resulting equation. If the same base is not possible, then take the logarithms of the two sides.

**Examples:**

- Solve: $2^x \times 4^{2x} = 32$

The equation can be written by expressing the exponentials in the same base as follows:

$2^x \times 2^{4x} = 32$   (Since $4^{2x} = (2^2)^{2x}$)

$2^{5x} = 2^5$

$5x = 5$   (Applying logarithms to both sides)

$x = 1$

- Solve: $\dfrac{3^x}{9^2} - 25 = 2$

$\dfrac{3^x}{3^4} = 27$

$3^{x-4} = 3^3$

$x - 4 = 3$

$x = 7$

- Solve $2^x = 24$

Here we cannot express 24 as an exact power of the base 2. Taking common logarithm of both sides, we get:

$x \log 2 = \log 24$

$x = \dfrac{\log 24}{\log 2} = \dfrac{1.3802}{0.3010} = 4.59$  (The Napierian Logarithm gives the same result)

With a sum of exponential functions we need to isolate the common functions and use substitutions for them as shown below before applying logarithm.

**Example:**

Solve: $2^{x+1} - 2^x = 5$

$2 \times 2^x - 2^x = 5$

Let $u = 2^x$

Therefore, substituting 'u' for $2^x$ we get,

$2u - u = 5$

$u = 5$

$2^x = 5$

$x\log 2 = \log 5$   (Taking logarithms)

$x = \dfrac{\log 5}{\log 2} = \dfrac{0.6990}{0.3010} = 2.32$

**EXERCISE 22.5**----------------------------------------------------------------------------------

Solve the following exponential equations:

1. $2^x = 8$    3. $6^x = 9$    5. $(9^x)^2 = 81$    7. $9.6^{2x} = 78$

2. $3^{2x} = 27$    4. $4^x = 13$    6. $(0.5)^y = 17$    8. $7.4^{2x-1} = 92$

----------------------------------------------------------------------------------

| Solutions: Exercise 22.5 | | | | | | | |
|---|---|---|---|---|---|---|---|
| **1.** 3 | **2.** 1.5 | **3.** 1.226 | **4.** 1.850 | **5.** 1 | **6.** −4.087 | **7.** 0.963 | **8.** 1.630 |

**EXERCISE 22.6**----------------------------------------------------------------------------------

Solve the following exponential equations:

1. $4^y = 64$

2. $3^x = \dfrac{1}{27}$

3. $8^{5x} = 256$

4. $(\dfrac{4}{3})^m = \dfrac{16}{9}$

5. $5^{-x} = \dfrac{1}{125}$

6. $(\dfrac{1}{3})^p = 81$

7. $6^{2x-1} = 1$

8. $3^{4x+3} = 25$

----------------------------------------------------------------------------------

| Solutions: Exercise 22.6 | | | |
|---|---|---|---|
| **1.** y = 3 | **3.** x = 0.533 | **5.** x = 3 | **7.** x = 0.5 |
| **2.** x = −3 | **4.** m = 2 | **6.** p = −4 | **8.** x = −0.0175 |

## LOGARITHMIC EQUATIONS

In order to solve a logarithmic equation we have to rewrite the equation in exponential form first. The logarithmic equation $\log_a b = x$ can be rewritten as $a^x = b$. Recall the following law of logarithm:

- $\log_a b = \dfrac{1}{\log_b a}$

- $\log_a x = \log_b x \times \log_a b$

- $\log_a x = \dfrac{\log_b x}{\log_b a}$

**Examples:**

- Solve: $\log_3 27 = y$

Rewriting in exponential form, we get:

$27 = 3^y$
$3^3 = 3^y$

Equating the powers on either side of the equation:

$y = 3$

- Solve: $\log x + \log 5 = 3$

This can be written as: $\log 5x = 3$

$5x = 10^3 = 1000$

$x = 200$

- Solve: $\log_2 x + \log_4 x = 12$

Here we have to express both logs to the same base, say base 2:

Therefore, $\log_2 x + \log_2 x \times \log_4 2 = 12$

$\log_2 x + \dfrac{1}{2}\log_2 x = 12$

$\dfrac{3}{2}\log_2 x = 12$

$\log_2 x = 8$

$x = 3$

- Solve: $\log_2 (x^2 - 6x + 21) = 4$

Using the definition of logarithm, we write

$x^2 - 6x + 21 = 2^4 = 16$

$x^2 - 6x + 5 = 0$

Factorising: $(x - 1)(x - 5) = 0$

$x = 1$ or $x = 5$

- Solve: $\ln(2x - 1) = 3$

Using the definition of logarithm, we write:

$2x - 1 = e^3$  (ln is used when we use 'e' as the base)

$2x = e^3 + 1$

$2x = 20.0855 + 1 = 21.0855$

$x = 10.54$

## EXERCISE 22.7-------------------------------------------------------------------

Solve the following logarithmic equations:

1.  $\log_3 x = 2$

2.  $\log_{1/2} x = 4$

3.  $\log_{1/3} 81 = m$

4.  $\log_a 10^{-3} = -3$

5.  $\log_p e^{-2} = -2$

6.  $\log x - \log 5 = 1$

7.  $\log_b 100 = 2$

8.  $\log_2 y + \log_2 8 = 6$

-------------------------------------------------------------------------------------

| Solutions: Exercise 22.7 | | | |
|---|---|---|---|
| **1.** $x = 9$ | **3.** $m = -4$ | **5.** $p = e = 2.71828$ | **7.** $b = 10$ |
| **2.** $x = \dfrac{1}{16}$ | **4.** $a = 10$ | **6.** $x = 50$ | **8.** $y = 8$ |

## EXERCISE 22.8-------------------------------------------------------------------

Solve the following logarithmic equations:

1.  $\log_4(x + 7) = 2$

2.  $\log_3(4x + 5) - \log_3(x + 2) = 1$

3.  $\log_4(x - 4) - \log_4 3 = 1/2$

4.  $\log_3 x + \log_9 x = 13.5$

5.  $\log_2 x^3 = 6$

6.  $\log_5(x^2 - 12x + 61) = 2$

7.  $\ln(4x - 1) = 2$

8.  $\ln(3x^2 + 5) = 3$

-------------------------------------------------------------------------------------

| Solutions: Exercise 22.8 | | | |
|---|---|---|---|
| **1.** $x = 9$ | **3.** $x = 10$ | **5.** $x = 4$ | **7.** $x = 2.097$ |
| **2.** $x = 1$ | **4.** $x = 3^9$ | **6.** $x = 6$ | **8.** $x = 2.242$ |

# CHAPTER 23

The **inequality signs** are:

- > (read as greater than)
- < (read as less than)
- ≥ (read as greater than or equal to)
- ≤ (read as less than or equal to)

In an equation, if one of the above signs replaces the equal sign it becomes an **inequation**.

Inequations are solved in a manner similar to equations. When we multiply an equation on both sides by a negative sign the equal sign remains as it is. However, in the case of inequations, the inequality symbol switches to the opposite symbol. That is, ">"becomes "<" and "<" becomes ">" when multiplied on both sides by a negative sign. Examine the following points:

- When $8 > 5$ is multiplied on both sides by a negative sign, we would write $-8 < -5$

- When $3 < 7$ is multiplied on both sides by a negative sign, we would write $-3 > -7$

- $6 > 4$ but the reciprocal of 6 is less than the reciprocal of 4; ie $\frac{1}{6} < \frac{1}{4}$

- $6 > -4$ and the reciprocal of 6 is also greater than the reciprocal of $-4$; ie $\frac{1}{6} > -\frac{1}{4}$

**Examples:**

- Solve: $x - 5 > 3$

To find the value of x we proceed as follows:

$x - 5 + 5 > 3 + 5$

$x > 8$  (ie the given inequality is true for any value of x greater than 8)

- Solve: $-x + 6 \geq 9$

$x - 6 \leq -9$   (Multiplying both sides by $-1$)
$x \leq -3$   (adding 6 to both sides)

- Solve: $\dfrac{2x+1}{x-2} > 1$

When solving a fractional inequality involving an algebraic expression in the denominator **it is necessary to multiply both sides by the square of the denominator.**

Multiplying both sides by $(x-2)^2$, we get

$(2x + 1)(x - 2) > (x - 2)^2$
$2x^2 - 3x - 2 > x^2 - 4x + 4$
$x^2 + x - 6 > 0$
$(x + 3)(x - 2) > 0$     (any value of x between –3 and 2 does not satisfy the inequality)
$x > 2$ or $x < -3$

(Decide the solution by checking whether the original inequality holds for $x > 2$, $x < 2$ and for $x < -3$ , $x > -3$)

Inequalities are represented on a **number line** as follows:

- $x > a$

- $x < a$

- $x \geq a$

- $x \leq a$

(Note the hollow and filled in circles)

If $x > a$ and $x < b$, both conditions can be combined and written as follows:

$b > x > a$ or $a < x < b$

- Solve: $|4x + 3| < 7$

This means either $4x + 3 < 7$ or $-4x - 3 < 7$ ie $-7 < 4x + 3$

$-7 < 4x + 3 < 7$
$-10 < 4x < 4$
$-2.5 < x < 1$

In other words, x lies between –2.5 and 1

We write this as an interval using parenthesis as (–2.5, 1)

**EXERCISE 23.1**------------------------------------------------------------------

Solve the following inequalities and show the solutions on a number line:

1. $3x - 2 > 7$      4. $3a - 5 \geq 4$      7. $2c - 7 \leq 5c + 2$

2. $4x + 3 < 9$      5. $6p + 8 < 7p + 5$      8. $-2m + 5 < 1 + 6m$

3. $2x + 1 \geq 5$      6. $12 - 3x \geq -16 + 4x$

------------------------------------------------------------------

## Solutions: Exercise 23.1

**1.** $x > 3$    **3.** $x \geq 2$    **5.** $p > 3$    **7.** $c \geq -3$

**2.** $x < 1.5$    **4.** $a \geq 3$    **6.** $x \leq 4$    **8.** $m > 0.5$

**EXERCISE 23.2**------------------------------------------------------------------

Solve and show solutions on a number line

1. $\dfrac{2x - 3}{6} < 1$      5. $\dfrac{2y + 2}{5} \geq \dfrac{4y + 3}{8}$

2. $\dfrac{-7 - 3c}{4} \leq 5$      6. $\dfrac{6}{b - 1} > 3$

3. $\dfrac{2p - 1}{7} > \dfrac{2p + 9}{3}$      7. $\dfrac{4}{a - 2} > 2$

4. $\dfrac{2b - 1}{4} < \dfrac{2b + 3}{3}$      8. $\dfrac{2x}{x - 2} < 3$

------------------------------------------------------------------

| | | | |
|---|---|---|---|
| **1.** $x < 4.5$ | **3.** $p < -8.25$ | **5.** $y \leq 0.25$ | **7.** $2 < a < 4$ |
| **2.** $c \geq -9$ | **4.** $b > -7.5$ | **6.** $1 < b < 3$ | **8.** $2 > x > 6$ |

## EXERCISE 23.3 --------------------------------------------------------------------------------

Solve and show solutions on a number line

1. $-2 < \dfrac{1}{y+1} < 2$

2. $\dfrac{2m}{m-2} \leq 1$

3. $\dfrac{2x}{x+2} > 1$

4. $\dfrac{c-2}{c+2} > 3$

5. $|3a - 4| > 5$

6. $|-4x + 3| > -3$

7. $9 + 4y > |3 - 2y|$

8. $9 - p \leq |7 - 2p|$

--------------------------------------------------------------------------------

**Solutions: Exercise 23.3**

| | |
|---|---|
| **1.** $-0.5 < y < -1.5$ | **5.** $-0.33 > a > 3$ |
| **2.** $-2 \leq m \leq 2$ | **6.** true for all values of x |
| **3.** $-2 < x < 2$ | **7.** $y > -1$ |
| **4.** $-2 > c > -4$ | **8.** $-2 \geq p \geq 5.33$ |

# CHAPTER 24

## CHANGING THE SUBJECT OF A FORMULA

A formula is an equation containing one or more variables.

**Examples:**

- $ax + b = c$

- $2abc + 3x = 2a$

- $a(bx + c) = d$

- $v^2 = u^2 + 2as$

- $\dfrac{1}{u} + \dfrac{1}{v} = \dfrac{1}{f}$

- $\dfrac{F - 32}{9} = \dfrac{c}{5}$

For each of the above, we could make one of the variables to be on the left side of the equation and the rest to be on the right side. The variable on the left side is then called the subject of the formula. Let us now take each one of the six formulae above in turn and make a particular variable the subject.

**Examples:**

- Make 'x' the subject of: $ax + b = c$

$ax + b = c$
$ax + b - b = c - b$
$ax = c - b$

Dividing both sides by "a", we get

$$x = \frac{c - b}{a}$$

- Make 'a' the subject of: $2abc + 3x = 2a$

$2abc + 3x = 2a$
$2a = 2abc + 3x$
$2a - 2abc = 2abc + 3x - 2abc$
$2a(1 - bc) = 3x$

Dividing both sides by $2(1 - bc)$, we get

$$a = \frac{3x}{2(1 - bc)}$$

- Make 'x' the subject of:  $a(bx + c) = d$;

$a(bx + c) = d$

Dividing both sides by 'a', we get

$bx + c = \dfrac{d}{a}$

$bx + c - c = \dfrac{d}{a} - c$

$bx = \dfrac{d - ac}{a}$

$x = \dfrac{d - ac}{ab}$

- Make 'u' the subject of:  $v^2 = u^2 + 2as$

$v^2 = u^2 + 2as$

Subtracting 2as from both sides, we get

$v^2 - 2as = u^2 + 2as - 2as$
$v^2 - 2as = u^2$

$u = \sqrt{v^2 - 2as}$

- Make 'v' the subject of:  $\dfrac{1}{u} + \dfrac{1}{v} = \dfrac{1}{f}$

$\dfrac{1}{u} + \dfrac{1}{v} = \dfrac{1}{f}$

Multiplying by the LCM of the denominators uvf we get:

$vf + uf = uv$
$vf + uf - vf = uv - vf$
$uf = uv - vf$
$uf = v(u - f)$
$v(u - f) = uf$

$v = \dfrac{uf}{u - f}$

- Make 'F' the subject of $\dfrac{F-32}{9} = \dfrac{C}{5}$

$$\dfrac{F-32}{9} = \dfrac{C}{5}$$

Cross multiplying we get:

$5(F-32) = 9C$
$5F - 160 = 9C$
$5F - 160 + 160 = 9C + 16$
$5F = 9C + 160$

$$F = \dfrac{9C + 160}{5}$$

## EXERCISE 24.1

Make 'x' the subject of the following formulae:

1. $x + y = z$
2. $2x - 3a = x + b$
3. $ax - b = c$
4. $lx - m = an$
5. $b = a + xt$
6. $xyz + 5x = 3y$
7. $x + y = 2x - z$
8. $ax + t = 3t - bx$

---

### Solutions: Exercise 24.1

| | | | |
|---|---|---|---|
| **1.** $x = z - y$ | **3.** $x = \dfrac{b+c}{a}$ | **5.** $x = \dfrac{b-a}{t}$ | **7.** $x = y + z$ |
| **2.** $x = 3a + b$ | **4.** $x = \dfrac{m+an}{l}$ | **6.** $x = \dfrac{3y}{yz+5}$ | **8.** $x = \dfrac{2t}{a+b}$ |

## EXERCISE 24.2

Make the letter in brackets on the right the subject of the formula:

1. $A = 2\pi rh$      (h)
2. $A = \dfrac{1}{2}(a + b)h$      (b)
3. $I = \dfrac{PNR}{100}$      (R)
4. $l = a + (n - 1)d$      (d)
5. $V = C(R + r)$      (R)
6. $s = \dfrac{u+v}{2}t$      (v)
7. $A = 2\pi r(r + h)$      (h)
8. $Pt = m(v - u)$      (u)

---

| | | | |
|---|---|---|---|
| **1.** $h = \dfrac{A}{2\pi r}$ | **3.** $R = \dfrac{100I}{PN}$ | **5.** $R = \dfrac{V}{C} - r$ | **7.** $h = \dfrac{A}{2\pi r} - r$ |
| **2.** $b = \dfrac{2A}{h} - a$ | **4.** $d = \dfrac{I-a}{n-1}$ | **6.** $v = \dfrac{2s}{t} - u$ | **8.** $u = v - \dfrac{Pt}{m}$ |

## EXERCISE 24.3-----------------------------------------------------------------------------

1. Change the subject of the formula $M = VD$ to $V$. If $M = 200$, $D = 8$, what is the value of $V$?

2. Make $N$ the subject of the formula $I = \dfrac{PNR}{100}$. If $I = 30$, $P = 500$ and $R = 3$, what is the value of $N$?

3. Rewrite the formula $v^2 = u^2 + 2as$ with $u$ as the subject. Calculate $u$ when $v = 44$, $a = 8$ and $s = 96$.

4. Make $r$ the subject of the formula $S = \dfrac{a}{1-r}$. Evaluate $r$ when $a = 4$ and $S = 8$

5. Make '$u$' the subject of: $\dfrac{1}{u} + \dfrac{1}{v} = \dfrac{1}{f}$ and calculate its value when $v = 20$ and $f = 60$.

6. Calculate the value of '$h$' in the formula: $A = 6r(r + h)$ when $r = 2.5$ and $A = 225$.

7. $A$, $P$ and $y$ are connected by the equation: $A = P + 2y^2t$. Calculate '$t$' if $A = 750$, $P = 550$ and $y = 4$.

8. If $P = 21$, $m = 2.5$ and $u = 36.6$ what is the value of $v$ given the formula $P = m(v - u)$?

-----------------------------------------------------------------------------

| | |
|---|---|
| **1.** $V = \dfrac{M}{D}$; $V = 25$ | **5.** $u = \dfrac{vf}{v-f}$; $u = -30$ |
| **2.** $N = \dfrac{100I}{PR}$; $N = 2$ | **6.** $h = \dfrac{A}{6r} - r$; $h = 12.5$ |
| **3.** $u = \sqrt{v^2 - 2as}$; $u = 20$ | **7.** $t = (A - P)/2y^2$; $6\dfrac{1}{4}$ |
| **4.** $r = 1 - \dfrac{a}{S}$; $r = \dfrac{1}{2}$ | **8.** $v = \dfrac{P}{m} + u$; $v = 45$ |

# CHAPTER 25

## SIMULTANEOUS EQUATIONS

For a linear equation with only one unknown variable we need only one equation to get a solution. When two variables are involved in an equation we need to have another equation to find a solution. Since we consider the two equations at the same time we refer to them as **simultaneous equations**.

Simultaneous equations are solved algebraically by two possible methods:

1.  **Substitution method**
2.  **Elimination method**

## SUBSTITUTION METHOD

In this method, we first obtain the value of 'y' in terms of 'x' (or 'x' in terms of 'y') from one of the two equations. We then substitute the value found in the other equation as shown in the examples that follow:

**Examples:**

• Solve the following simultaneous equations using the substitution method:

$x + y = 7$
$x - y = 1$

Name the equations as follows:

$x + y = 7$ -------------------(1)
$x - y = 1$ -------------------(2)

From equation (1), $x = 7 - y$ ----(3)

Substituting the value of x in equation (2), we have:

$7 - y - y = 1$
$7 - 2y = 1$
$-2y = -6$   (Subtracting 7 from both sides)
$y = 3$   (Dividing both sides of the equation by $-2$)

Therefore, $x = 7 - 3 = 4$   [obtained by substituting in equation (3)]

The solution to the simultaneous equations is: $x = 4$, $y = 3$

- Solve the following simultaneous equations using the substitution method:

x + 5y = 10
3x + y = 58

Name the equations as follows:

x + 5y = 10 --------------------(1)
3x + y = 58 --------------------(2)

From equation (1), x = 10 − 5y ---- (3)

Substituting the value of x in equation (2), we have:

3(10 − 5y) + y = 58
30 − 15 y + y = 58    (Expanding the brackets)
30 − 14y = 58
30 − 14y − 30 = 58 − 30    (Subtracting 30 from both sides)
−14y = 28
y = −2    (Dividing both sides of the equation by −14)

Therefore, x = 10 − (−10) = 20

The solution to the simultaneous equations is:  x = 20, y = −2

- Solve the following simultaneous equations using the substitution method:

2x −7y = −17
3x + 5y = 21

Name the equations as follows:

2x − 7y = −17 ----------------(1)
3x + 5y = 21 ---------------(2)

From equation (1), x = $\dfrac{7y-17}{2}$ ---- (3)

Substituting the value of x in equation (2), we have:

$3(\dfrac{7y-17}{2}) + 5y = 21$

3(7y − 17) + 10y = 42    (Multiplying by 2)
21y − 51 + 10y = 42
31y = 93
y = 3

Therefore, x = $\dfrac{21-17}{2}$ = 2

The solution to the simultaneous equations is:  x = 2, y = 3

- Solve the following simultaneous equations using the substitution method:

$x + y = 8$

$\dfrac{x}{5} + \dfrac{y}{3} = 2$

Name the equations as follows:

$x + y = 8$ -----------------(1)

$\dfrac{x}{5} + \dfrac{y}{3} = 2$ ---------------(2)

Multiply equation (2) by 15 (the LCM of the denominators 5 and 3) to get a new equation and call it equation (3)

$3x + 5y = 30$ -------------(3)

From equation (1), $x = 8 - y$

Substituting the value of x in equation (3), we have:

$3(8 - y) + 5y = 30$
$24 - 3y + 5y = 30$
$2y = 6$
$y = 3$

Therefore, $x = 8 - 3 = 5$

The solution to the simultaneous equations is: $x = 5, y = 3$

- Solve the following simultaneous equations using the substitution method:

$\dfrac{x}{4} - \dfrac{y}{8} = 2$

$\dfrac{x}{5} + \dfrac{y}{2} = 4$

Name the equations as follows:

$\dfrac{x}{4} - \dfrac{y}{8} = 2$ ----------------(1)

$\dfrac{x}{5} + \dfrac{y}{2} = 4$ -------------- (2)

Multiplying equation (1) by 8 and equation (2) by 10, we have

$2x - y = 16$ --------------------(3)

$2x + 5y = 40$ ---------------(4)

From equation (3), $y = 2x - 16$

Substituting for y in (4):

$2x + 5(2x - 16) = 40$
$2x + 10x - 80 = 40$
$12x = 120$
$x = 10$

Therefore, $y = 20 - 16 = 4$

The solution to the simultaneous equations is:  $x = 10, y = 4$

## ELIMINATION METHOD

In this method we try and make either the absolute value of the coefficient of x or y equal in both equations (unless they are already equal) and eliminate one of the variables by addition or subtraction of the two equations.  This is illustrated below with three of the same examples used in the substitution method.

**Examples:**

- Solve the following simultaneous equation using the Elimination method:

$x + y = 7$
$x - y = 1$

Name the equations as follows:

$x + y = 7$ --------------------(1)
$x - y = 1$ --------------------(2)

The absolute values of the coefficients of both x and y are already equal in both equations.

Observe that 'y' can be eliminated if equations (1) and (2) are added.

$2x = 8$   $[(1) + (2)]$
$x = 4$

Observe again that 'x' can be eliminated if equation (2) is subtracted from equation (1)

$2y = 6$   [(1) − (2)]
$y = 3$

The solution to the simultaneous equations is:  $x = 4, y = 3$

- Solve the following simultaneous equations using the Elimination method:

$x + 5y = 10$
$3x + y = 58$

Name the equations as follows:

$x + 5y = 10$ --------------------(1)
$3x + y = 58$ --------------------(2)

Multiply equation (1) by 3 to make the absolute coefficient of x to be the same as in equation (2) and call it equation (3)

$(1) × 3 = 3x + 15y = 30$ -----------(3)

Subtract equation (2) from equation (3) to give:

$14y = -28$   [(3) − (2)]
$y = -2$

From equation (1), $x = 10 - 5y$

Substituting −2 for y: $x = 10 - 5(-2) = 20$

The solution to the simultaneous equations is:  $x = 20, y = -2$

- Solve the following simultaneous equations using the elimination method:

$2x - 7y = -17$
$3x + 5y = 21$

Name the equations as follows:

$2x - 7y = -17$ ----------------(1)
$3x + 5y = 21$ ----------------(2)

Now we need to have the same absolute coefficient of x or y in both equations. Since x is the simpler option, multiply equation (1) by 3 (the coefficient of x in the second equation) and multiply equation (2) by 2 (the coefficient of x in the first equation).

$6x - 21y = -51$ ---------(3)   [(1) × 3]

$6x + 10y = 42$ -----------(4)   [(2) × 2]

Subtract equation (4) from equation (3) to get:

$-31y = -93$
$y = 3$

From equation (1), $x = \dfrac{7y - 17}{2}$

Substituting 3 for y, $x = \dfrac{21 - 17}{2} = 2$

The solution to the simultaneous equations is: $x = 2, y = 3$

## EXERCISE 25.1----------------------------------------------------------------------------

Solve the following simultaneous equations using both the method of substitution and the method of elimination

1. $x - y = 5$
   $x + y = 19$

2. $x + y = 2$
   $x - y = -8$

3. $x + y = -14$
   $x - y = -2$

4. $x + y = 5$
   $2x - y = 1$

5. $2x + 3y = 25$
   $x - y = -5$

6. $5x + 7y = 47$
   $4x + y = 10$

7. $2x - y = 8$
   $10x - 9y = 48$

8. $x + 3y = 25$
   $3x - y = 25$

------------------------------------------------------------------------------------------------

| Solutions: Exercise 25.1 | | | | | | | |
|---|---|---|---|---|---|---|---|
| **1.** x =12 | y = 7 | **3.** x = –8 | y = –6 | **5.** x = 2 | y = 7 | **7.** x = 3 | y = –2 |
| **2.** x = –3 | y = 5 | **4.** x = 2 | y = 3 | **6.** x = 1 | y = 6 | **8.** x = 10 | y = 5 |

## EXERCISE 25.2------------------------------------------------------------

Solve the following simultaneous equations using the method of your choice.

1. $5x + 4y = 7$
   $4x + 5y = 2$

2. $x + 5y = 18$
   $3x + 2y = 41$

3. $2x + 3y = -3$
   $7x - 4y = -25$

4. $5x - 6y = -11$
   $4x - 7y = -33$

5. $6x - 3y = 15$
   $8x - 5y = 17$

6. $3x - 2 = 2y$
   $2x + 10y = 7$

7. $9y = 3 + 8x$
   $8y - 6x = 3$

8. $11x = 10y - 35$
   $3x = -5y + 60$

### Solutions: Exercise 25.2

| | | | | | | | |
|---|---|---|---|---|---|---|---|
| **1.** $x = 3$ | $y = -2$ | **3.** $x = -3$ | $y = 1$ | **5.** $x = 4$ | $y = 3$ | **7.** $x = 0.3$ | $y = 0.6$ |
| **2.** $x = 13$ | $y = 1$ | **4.** $x = 11$ | $y = 11$ | **6.** $x = 1$ | $y = 1/2$ | **8.** $x = 5$ | $y = 9$ |

## EXERCISE 25.3------------------------------------------------------------

Solve the following simultaneous equations:

1. $2x - y = 4$
   $\dfrac{x}{2} + \dfrac{y}{4} = 5$

2. $4x - y = 3$
   $\dfrac{x}{3} + \dfrac{2y}{3} = 1$

3. $x - 3y = -2$
   $x - \dfrac{9y}{4} = 1\dfrac{3}{4}$

4. $\dfrac{x}{3} + \dfrac{y}{4} = 4$
   $\dfrac{x}{6} - \dfrac{y}{9} = \dfrac{1}{9}$

5. $\dfrac{x}{8} - \dfrac{y}{9} = 1$
   $\dfrac{x}{4} + \dfrac{y}{3} = 7$

6. $\dfrac{x}{2} + \dfrac{y}{3} = 4$
   $x - \dfrac{2y}{3} = 0$

7. $\dfrac{x}{3} - \dfrac{y}{2} = \dfrac{1}{6}$
   $\dfrac{x}{6} + \dfrac{y}{2} = \dfrac{5}{6}$

8. $\dfrac{x}{2} + \dfrac{y}{3} = 2\dfrac{2}{3}$
   $\dfrac{x}{3} + \dfrac{y}{2} = 1\dfrac{1}{2}$

### Solutions: Exercise 25.3

| | | | | | | | |
|---|---|---|---|---|---|---|---|
| **1.** $x = 6$ | $y = 8$ | **3.** $x = 13$ | $y = 5$ | **5.** $x = 16$ | $y = 9$ | **7.** $x = 2$ | $y = 1$ |
| **2.** $x = 1$ | $y = 1$ | **4.** $x = 6$ | $y = 8$ | **6.** $x = 4$ | $y = 6$ | **8.** $x = 6$ | $y = -1$ |

1.  The sum of two numbers is 43 and their difference is 13. What are the two numbers?

2.  One quarter of the sum of two numbers is their difference. Three times the bigger number is 21 less than six times the smaller number. Find the numbers.

3.  When a number is divided by another number, the quotient is 5 and the remainder is 3. Six times the divisor is 4 greater than the dividend. What are the two numbers?

4.  A man invests $80000 in government bonds yielding 4.25% return and in shares yielding 6% return. The investments are made in units of $100 each. If he desires an annual income of $3855, how many units of bonds and shares should he buy?

5.  The cost of 6 chairs and 4 tables is $1070. The cost of 8 similar chairs and 6 similar tables is $1520. Determine the cost per chair and table.

6.  620 tickets in total were sold for a concert. Adult tickets cost $20 and children's cost $8. The total collection was $9760. How many tickets of each kind were sold?

7.  A man swims a distance of 12km upstream in 3 hours. He swims the same distance downstream in 1.2 hours. What are the speeds of the man and the current?

8.  A rectangular field has a perimeter of 48 metres. The length of the field is 6 metres more than the width. Determine the length and width of the field.

-----------------------------------------------------------------------------------------

| Solutions: Exercise 25.4 | | | |
|---|---|---|---|
| **1.** 28 and 15 | **3.** 38 and 7 | **5.** Chair = $85<br>Table = $140 | **7.** Man = 7km/hr<br>Current = 3km/hr |
| **2.** 35 and 21 | **4.** 540 units of bonds<br>260 units of shares | **6.** Adult = 400<br>Children = 220 | **8.** Length = 15m<br>Width = 9m |

Algebraic expressions that are made up of more than one term are called polynomial functions. The general form of a polynomial function in one variable x is written as:

$f(x) = ax^2 + bx + c$, where a is the coefficient of $x^2$, b is the coefficient of x and c is a constant term.

An equation involving a polynomial expression containing a single variable with the highest power of 2 is called a quadratic (polynomial) equation.

The general form of a **quadratic equation** is thus: $ax^2 + bx + c = 0$

A quadratic equation has two values which satisfy it. That is, when the values are substituted for x, the equation reduces to zero. The two values can be real numbers and different, real and equal or complex. The values of x which satisfy the equation are referred to as the **roots** of the equation.

**Examples:**

- Solve: $x^2 - 3x + 2 = 0$

The equation can be factorised and stated as $(x - 1)(x - 2) = 0$. The roots are obtained by separately equating $(x - 1)$ and $(x - 2)$ to zero.

$(x - 1) = 0$     $(x = 1)$
$(x - 2) = 0$     $(x = 2)$

The roots of the equation are therefore 1 and 2 (real but different).

- Solve: $6x^2 + x - 2 = 0$

$(2x - 1)(3x + 2) = 0$

The roots are: $\dfrac{1}{2}$ and $-\dfrac{2}{3}$

- Solve: $x^2 - 6x + 9 = 0$

Factorising we get $(x - 3)(x - 3) = 0$. The roots are 3 and 3 (real and equal).

- Solve $4x^2 + 12x + 9 = 0$

$(2x + 3)(2x + 3) = 0$

The two equal roots are: $-\dfrac{3}{2}$ and $-\dfrac{3}{2}$

Given the two roots, we can determine the quadratic equation. If the roots are 1 and 2 then the quadratic equation is $(x - 1)(x - 2) = 0$. Expanding the brackets we get $x^2 - 3x + 2 = 0$. If the roots are 3 and 3, then the quadratic equation is $(x - 3)(x - 3) = 0$. Expanding the brackets we get $x^2 - 6x + 9 = 0$.

In general, if $\alpha$ and $\beta$ are the roots of a quadratic equation, then the equation is written as:

$(x - \alpha)(x - \beta) = 0$
$x^2 - \alpha x - \beta x + \alpha\beta = 0$
$x^2 - (\alpha + \beta)x + \alpha\beta = 0$

A quadratic equation is read as $x^2 -$ (sum of roots)$x +$ product of roots $= 0$

Comparing this with the general equation: $ax^2 + bx + c = 0$ we have:

$$x^2 - (\alpha + \beta)x + \alpha\beta = 0 \quad \text{and} \quad x^2 + \frac{b}{a}x + \frac{c}{a} = 0$$

Notice that: $\alpha + \beta = -\dfrac{b}{a}$ and $\alpha\beta = \dfrac{c}{a}$

- The sum of the roots of a quadratic equation $= \dfrac{-b}{a}$

- The product of the roots of a quadratic equation $= \dfrac{c}{a}$

**Examples:**

- The quadratic equation $x^2 - 5x + 6 = 0$ has two roots whose sum is 5 and product is 6.

- The quadratic equation $3x^2 + 7x - 4 = 0$ has roots whose sum is $-\dfrac{7}{3}$ and product $-\dfrac{4}{3}$.

## COMPLETING THE SQUARE & THE QUADRATIC FORMULA

The general quadratic equation is $ax^2 + bx + c = 0$. Dividing throughout by the coefficient of $x^2$, "a" we get:

$$x^2 + \frac{b}{a}x + \frac{c}{a} = 0 \text{ or } x^2 + \frac{b}{a}x = -\frac{c}{a}$$

Half the coefficient of x is $\dfrac{b}{2a}$. We add the square of this to both sides of the above equation:

$$x^2 + \frac{b}{a}x + \left(\frac{b}{2a}\right)^2 = -\frac{c}{a} + \left(\frac{b}{2a}\right)^2$$

This simplifies to:

$$\left(x + \frac{b}{2a}\right)^2 = \frac{b^2 - 4ac}{2a} \quad \text{(notice the completed square on the left)}$$

Taking square root on both sides:

$$x + \frac{b}{2a} = \pm\frac{\sqrt{b^2 - 4ac}}{2a}$$

$$x = \frac{-b + \sqrt{b^2 - 4ac}}{2a} \quad \text{or} \quad \frac{-b - \sqrt{b^2 - 4ac}}{2a}$$

$$x = \frac{-b \pm \sqrt{b^2 - 4ac}}{2a} \quad \text{is called \textbf{the Quadratic Formula}}$$

'$b^2 - 4ac$' is called the **discriminant**. The value of this decides whether the roots are real, equal or complex.

- If the discriminant is **positive** the roots are **real** and different
- If the discriminant is **negative** the roots are **complex**
- If the discriminant is **zero** the roots are **equal**

**Examples:**

- Solve by completing the square:  $2x^2 + 4x - 15 = 0$

Divide by the coefficient of $x^2$ to get:

$$x^2 + 2x - \frac{15}{2} = 0$$
$$x^2 + 2x = \frac{15}{2}$$

Complete the square of the left hand side by taking half the coefficient of x as follows:
$$(x + 1)^2 = \frac{15}{2} + 1$$

Note that the square of (x + 1) has a "1" additional to $x^2 + 2x$ and so add '1' to the right hand side to maintain balance.

$(x + 1)^2 = 8.5$

Take square root on both sides to get:

$x + 1 = \pm\sqrt{8.5} = \pm 2.9154$
x = 2.9154 – 1 or –2.9152 – 1
x = 1.9154 or –3.9154
x = 1.92 or –3.92     (corrected to two decimal places)

- Solve:  $2x^2 + 3x - 14 = 0$ using the quadratic formula

a = 2, b = 3, c = –14

$$x = \frac{-b + \sqrt{b^2 - 4ac}}{2a} \quad or \quad \frac{-b - \sqrt{b^2 - 4ac}}{2a}$$

$$x = \frac{-3 + \sqrt{9 - 112}}{4} \quad or \quad \frac{-3 - \sqrt{9 - 112}}{4}$$

x = 2 or –3.5

- Solve $3x^2 - x - 5 = 0$,  giving the answer correct to 2 places of decimal

a = 3, b = –1, c = –5

$$x = \frac{-b + \sqrt{b^2 - 4ac}}{2a} \quad or \quad \frac{-b - \sqrt{b^2 - 4ac}}{2a}$$

$$x = \frac{1 + \sqrt{1 + 60}}{6} \quad or \quad \frac{1 - \sqrt{1 + 60}}{6}$$

$$x = \frac{1 + \sqrt{61}}{6} \quad or \quad \frac{1 - \sqrt{1 + 61}}{6}$$

$$x = \frac{8.8102}{6} \quad or \quad \frac{-6.8102}{6}$$

x = 1.30 or –1.14

**EXERCISE 26.1**--------------------------------------------------------------------------------

1. Write down the quadratic equation whose roots are 2 and 5.

2. Write down the quadratic equation the sum of whose roots is –3 and the product of the roots is 2.

3. If one root of a quadratic equation is 4 and the product of the roots is 2 write down the equation.

4. For what value of 'c' is y = 2 a root of the equation $c(y^2 + 6) = y(c + 6)$?

5. Use the identity $(\alpha - \beta)^2 = (\alpha + \beta)^2 - 4\alpha\beta$ to calculate the difference between the two roots of a quadratic equation, the sum of whose roots are –1 and the product of the roots is –12.

6. If one root of a quadratic equation $6x^2 - 19x + c = 0$ is $-\dfrac{1}{3}$, what is the other root?

7. For what value of 'c' will the quadratic equation $4x^2 - 36x + c = 0$ have equal roots? Find the roots.

8. Determine whether each of the following equations has real and different roots, real and equal roots or complex roots.

   (i) $x^2 - 2x - 2 = 0$       (ii) $2x^2 - 5x + 6 = 0$       (iii) $9x^2 - 6x + 1 = 0$

--------------------------------------------------------------------------------

| Solutions: Exercise 26.1 | | | |
|---|---|---|---|
| **1.** $x^2 - 7x + 10 = 0$ | **3.** $2x^2 - 9x + 4 = 0$ | **5.** 7 | **7.** c = 81, Roots = 4.5, 4.5 |
| **2.** $x^2 + 3x + 2 = 0$ | **4.** c = 1.5 | **6.** 3.5 | **8.** (i) real & different (ii) complex (iii) equal |

**EXERCISE 26.2**--------------------------------------------------------------------------------

Solve the following equations by the method of factorisation:

1. $x^2 - 4 = 0$

2. $x^2 + 15 = 79$

3. $2(x^2 - 6) = x^2 - 3$

4. $3(x^2 + 5) + 2(8 - x^2) = 35$

5. $x^2 + x - 2 = 0$

6. $x^2 = x + 6$

7. $6x^2 - 3 = 7x$

8. $5x^2 + 8 = 14x$

--------------------------------------------------------------------------------

| Solutions: Exercise 26.2 | | | |
|---|---|---|---|
| **1.** x = 2 or – 2 | **3.** 3 or – 3 | **5.** x = 1 or – 2 | **7.** x = 1.5 or –0.33 |
| **2.** x = 8 or – 8 | **4.** 2 or – 2 | **6.** x = 3 or – 2 | **8.** x = 2 or 0.8 |

Solve the following equations by the method of completing the square.

1. $x^2 - 5x - 6 = 0$      4. $6x^2 - 11x - 10 = 0$      7. $2x^2 + 15 = 13x$

2. $x^2 - 7x + 12 = 0$      5. $9x^2 + 15x + 4 = 0$      8. $63x^2 = 29x + 24$

3. $2x^2 - 7x + 3 = 0$      6. $4x^2 - 5x = 6$

**Solutions: Exercise 26.3**

| | | | |
|---|---|---|---|
| **1.** $x = 6$ or $-1$ | **3.** $x = 3$ or $\frac{1}{2}$ | **5.** $x = -\frac{1}{3}$ or $-1\frac{1}{3}$ | **7.** $x = 5$ or $1\frac{1}{2}$ |
| **2.** $x = 3$ or $4$ | **4.** $x = 2\frac{1}{2}$ or $-\frac{2}{3}$ | **6.** $x = 2$ or $-\frac{3}{4}$ | **8.** $x = \frac{8}{9}$ or $-\frac{3}{7}$ |

Solve the following using the quadratic formula giving the answer corrected to two decimal places where required:

1. $3m^2 + 8m + 1 = 0$      4. $5a^2 + 12a - 4 = 0$      7. $x^2 - 6x + 7 = 0$

2. $y^2 + 2y - 8 = 0$      5. $x^2 + 2 = -8x$      8. $4p^2 - 4p - 5 = 0$

3. $5x^2 - 4x = 28$      6. $2x^2 - x = 1$

**Solutions: Exercise 26.4**

| | | | |
|---|---|---|---|
| **1.** $m = -2.54$ or $-0.13$ | **3.** $x = 2.80$ or $-2$ | **5.** $x = -0.26$ or $-7.74$ | **7.** $x = 4.41$ or $1.59$ |
| **2.** $y = 2$ or $-4$ | **4.** $a = 0.3$ or $-2.70$ | **6.** $x = 1$ or $-0.5$ | **8.** $p = 1.72$ or $-0.72$ |

Solve the following using the quadratic formula giving the answer corrected to two decimal places.

1. $6p^2 - 23p + 21 = 0$      4. $8c^2 + 10c - 117 = 0$      7. $6c^2 - 43c + 70 = 0$

2. $21x^2 - 4x - 32 = 0$      5. $6r^2 - 19r - 36 = 0$      8. $8x^2 - 34x - 69 = 0$

3. $4m^2 - 28m + 33 = 0$      6. $35y^2 - 2y - 48 = 0$

**EXERCISE 26.6**-----------------------------------------------------------------------------------

1.  The length of a hall is 3 metres more than its breadth. If its area is 180 m$^2$, find its length.

2.  The diagonal of a square is 50 cm. What is the length of a side?

3.  In the formula h = ut + $\dfrac{1}{2}$gt$^2$, u = 50, h = 120 and g = 10. Calculate the value of t.

4.  In a right angled triangle the hypotenuse is twice as long as the shorter of the other two sides. The longer of the other two sides is 10cm. What is the length of the hypotenuse?

5.  A landlord rents 10 similar apartments for $500 per week each. He reckons that for each $50 increase in rent he will not be able to rent one apartment. If he is satisfied with a rent revenue per week of $3200, what should the rent per week be?

6.  Two men start to run from the same point. If one runs North at 10 km/hr and the other runs East at 12 km/hr, when will they be 8 km apart?

7.  The profit from the sale of a product is given by the formula P = (n$^2$/8) – 16n where n is the number of units of product sold. How many units were sold if the profit made was $7680?

8   The cost (C) of manufacture of a product by a company is given by C = 30x$^2$ – 1800 where x is the number of units manufactured. The total revenue (R) is given by the equation R = 120x. How many units must be manufactured and sold to break even (ie to make neither a profit nor a loss)?

-------------------------------------------------------------------------------------------

# CHAPTER 27

Any algebraic expression containing variables and/or a constant is called a **polynomial**. It may or may not contain powers of the variables. A polynomial with one variable is commonly encountered. The degree of the one variable polynomial is the largest power of the variable.

A polynomial function is usually represented by the notation $f(x)$ if it contains the variable x only. If the variable is say, t, then the polynomial function in t is represented by $f(t)$ and so on.

The degree of the polynomial in x, say, $f(x) = x^2 + 2x + 1$ is two. It is called a quadratic polynomial.

| The general form of a two degree polynomial is $f(x) = ax^2 + bx + c$ |
| --- |

The degree of the polynomial in t, $f(t) = t^3 + 3t^2 + 3t + 1$ is three. It is a cubic polynomial.

| The general form of a three degree polynomial is $f(x) = ax^3 + bx^2 + cx + d$ |
| --- |

A degree one polynomial is one where the highest power of the variable is 1, as shown below. This is also called a linear function or a linear binomial.

$f(m) = 3m + 4$

A degree zero polynomial is when the polynomial has a constant term only

$f(x) = 5 \quad (5 = 5x^0)$

An 'n' degree polynomial in x is written as:

$f(x) = a_nx^n + a_{n-1}x^{n-1} + a_{n-2}x^{x-2} + \ldots\ldots\ldots\ldots a_1x + a_0$

$a_n$ is called the leading term of the polynomial, and cannot be 0.

A monomial has only one term, a binomial has two terms, a trinomial has three terms and so on. All of these are referred to as polynomials.

## ROOTS OF A POLYNOMIAL

The roots (or zeros) of a polynomial function of a given variable are the values of the variable that will make the function equal zero. A polynomial of degree 'n' has 'n' roots which may be real or complex.

The largest number of real roots that an 'n' degree polynomial could have is n. The smallest number is 1, real or complex. If a polynomial can be factorised into say three linear factors (x – a), (x – b) and (x – c) we can write:  f(x) = (x – a)(x – b)(x – c).  This means that a, b and c are the roots of f(x).

All roots of a polynomial function need not be integers; there can be decimal roots as well as complex roots.

## DIVISION OF POLYNOMIALS

When a polynomial is divided by a linear factor the quotient will be a polynomial of a lower degree and a remainder which may or may not be a constant.  In some cases there may not be any remainder.

The same is true when a polynomial is divided by another polynomial with a lower degree.

**Examples:**

- Divide $x^3 - 6x - 4$ by $x - 2$

$$
\begin{array}{r}
x^2 \quad +2x \quad -2 \qquad \text{(Polynomial a lower degree than the dividend)} \\
x - 2 \,\big|\, \overline{x^3 \qquad\quad -6x \quad -4\phantom{0}} \\
\underline{x^3 \;-2x^2} \\
+2x^2 \;-6x \\
\underline{+2x^2 \;-4x} \\
-2x \quad -4 \\
\underline{-2x \quad +4} \\
-8 \qquad \text{(Remainder is a constant)}
\end{array}
$$

- Divide $p^3 - 12p - 16$ by $p^2 - 2p + 8$

$$
\begin{array}{r}
p \qquad +2 \\
p^2 - 2p + 8 \,\big|\, \overline{p^3 \qquad\qquad -12p \quad -16} \\
\underline{p^3 \;-2p^2 \;+8p} \\
2p^2 \;-20p \;-16 \\
\underline{2p^2 \;-4p \;+16} \\
-16p \;-32 \qquad \text{(Remainder not a constant)}
\end{array}
$$

- Divide $m^4 + 3m^3 + 7m^2 + 12m + 12$ by $m^2 + 3m + 3$

$$
\begin{array}{r}
m^2 \qquad\qquad +4 \\
m^2 + 3m + 3 \,\big|\, \overline{m^4 \;+3m^3 \;+7m^2 \;+12m \;+12} \\
\underline{m^4 \;+3m^3 \;+3m^2} \\
+4m^2 \;+12m \;+12 \\
\underline{+4m^2 \;+12m \;+12} \\
0 \qquad \text{(no remainder)}
\end{array}
$$

The remainder states that if a polynomial $f(x)$ is divided by $(x - a)$, the remainder is obtained by evaluating $f(a)$. If we call the quotient $P(x)$, then we can write $f(x) = (x - a)P(x) + f(a)$.

**Examples:**

- Use the remainder theorem to find the remainder when $2x^2 + 5x - 9$ is divided by $x - 2$. Check the answer by long division.

$f(x) = 2x^2 + 5x - 9$

Remainder $= f(2) = 2(2)^2 + 5(2) - 9 = 8 + 10 - 9 = 9$

Long Division:

$$\begin{array}{r} 2x \phantom{xx} +9 \phantom{xxx} \\ x - 2 \overline{\smash{\big)}\, 2x^2 \ +5x \phantom{xx} -9} \\ \underline{2x^2 \ -4x \phantom{xxxxxx}} \\ 9x \phantom{x} -9 \\ \underline{9x \ -18} \\ 9 \end{array}$$

Therefore the quotient is $2x + 9$ and the remainder is 9.

- Find the remainder when $3x^3 + 2x^2 - 5x + 7$ is divided by $(x + 3)$

Remainder $= f(-3) = 3(-3)^3 + 2(-3)^2 - 5(-3) + 7$

$= -81 + 18 + 15 + 7 = -41$

Long Division:

$$\begin{array}{r} 3x^2 \phantom{xx} -7x \phantom{xxx} +16 \phantom{xxx} \\ x + 3 \overline{\smash{\big)}\, 3x^3 \ +2x^2 \phantom{xx} -5x \phantom{xxx} +7} \\ \underline{3x^3 \ +9x^2 \phantom{xxxxxxxxxxx}} \\ -7x^2 \phantom{xx} -5x \phantom{xxxxxx} \\ \underline{-7x^2 \ -21x \phantom{xxxxxx}} \\ +16x \phantom{xx} +7 \\ \underline{-16x \ +48} \\ -41 \end{array}$$

Therefore the quotient is $3x^2 - 7x + 16$ and the remainder is $-41$.

This theorem states that if a polynomial $f(x)$ is divided by $(x - a)$ and the remainder is zero, then $(x - a)$ is a factor of $f(x)$. In general, $f(x) = (x - a)P(x)$.

**Examples:**

- Use the factor theorem to factorise $f(x) = x^2 + x - 2$

The coefficient of $x^2$ is 1 and the constant term is $-2$. Their product is $-2$. The integer factors of $-2$ are 1, $-1$, 2 and $-2$.

There may be fractional roots of $f(x)$. We shall find these, if any, by long division. We try the integer factors of $-2$ in turn to find out which ones make $f(x) = 0$

$f(1) = 1 + 1 - 2 = 0$
$f(-1) = 1 - 1 - 2 = $ not zero
$f(2) = 4 + 2 - 2 = $ not zero
$f(-2) = 4 - 2 - 2 = 0$

Therefore, $x = 1$ and $x = -2$ are roots. The factors are therefore $(x - 1)$ and $(x + 2)$

Hence, $f(x) = (x - 1)(x + 2)$

- Use the factor theorem and long division to find the factors of
  $f(x) = 3x^2 + 2x - 1$

The coefficient of $x^2$ is 3 and the constant term is $-1$. Their product is $-3$. The factors of $-3$ are 1, $-1$, 3 and $-3$.

$f(1) = 3 + 2 - 1 = $ not zero
$f(-1) = 3 - 2 - 1 = 0$
$f(3) = 27 + 6 - 1 = $ not zero
$f(-3) = 27 - 6 - 1 = $ not zero

Therefore, $x = -1$ is a root and $x + 1$ is a factor of $f(x)$.

A quadratic equation must have two roots. The other root can be found by long division as follows:

$$
\begin{array}{r}
3x \quad -1 \\
x + 1 \overline{\smash{\big)}\ 3x^2\ +2x\ -1} \\
\underline{3x^2\ +3x\phantom{\ -1}} \\
-x\ -1 \\
\underline{-x\ -1} \\
0
\end{array}
$$

The other factor of $f(x)$ is $3x - 1$. Therefore, $f(x) = (x - 1)(3x - 1)$

- Factorise: $f(x) = x^3 - 6x^2 + 7x + 6$

The coefficient of $x^3$ is 1 and the constant term is 6. Their product is 6
The factors of 6 are 1, 2, 3 and 6.

We try these in turn to find out which ones make $f(x) = 0$

$f(1) = (1)3 - 6(1)^2 + 7(1) + 6 = $ not zero
$f(2) = (2)^3 - 6(2)^2 + 7(2) + 6 = $ not zero
$f(3) = (3)^3 - 6(3)^2 + 7(3) + 6 = $ zero
$f(6) = (6)^3 - 6(6)^2 + 7(6) + 6 = $ not zero

Therefore, $x = 3$ is one root and $x - 3$ is a factor. We now do the long division to find the other factor.

$$
\begin{array}{r}
x^2 \quad -3x \quad -2 \\
x - 3 \enclose{longdiv}{x^3 \quad -6x^2 \quad +7x \quad +6} \\
\underline{x^3 \quad -3x^2} \\
-3x^2 \quad +7x \\
\underline{-3x^2 \quad +9x} \\
-2x \quad +6 \\
\underline{-2x \quad +6} \\
0
\end{array}
$$

Hence, the factors of $f(x) = x^3 - 6x^2 + 7x + 6$ are $(x - 3)$ and $(x^2 - 3x - 2)$,
Note that one is a linear factor and the other is a quadratic factor.

## GRAPHS OF SOME POLYNOMIAL FUNCTIONS

Linear functions:    $f(x) = ax + b$

Linear (1st Degree) Increasing Function with a
**positive gradient** and a **negative y-intercept.**

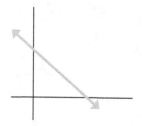

Linear (1st Degree) Decreasing Function with a
**negative gradient** and a **positive y-intercept.**

**Examples:**

- $f(x) = 2x - 3$

Increasing function with a gradient of 2 and y-intercept of –3

- $f(x) = -\dfrac{1}{2}x + 4$

Decreasing function with a gradient of $-\dfrac{1}{2}$ and a y-intercept of + 4

---

## QUADRATIC FUNCTIONS

Parabolic (2nd Degree) concave graph showing the turning point that gives the **minimum value of the function.**

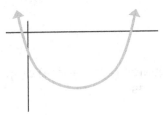

Parabolic (2nd Degree) convex graph showing the turning point that gives the **maximum value of the function.**

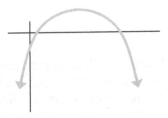

For the quadratic function: $f(x) = ax^2 + bx + c$, the x-coordinate of the turning point is given by $\dfrac{-b}{2a}$. A positive value of 'a' means the graph is concave and a negative means the curve is convex.

**Example:**

- Find the coordinates of the turning point of: $f(x) = 2x^2 + 3x - 2$ and state whether it is maximum or minimum

$$x - \text{coordinate of the turning point} = \dfrac{-b}{2a} = \dfrac{-3}{4}$$

$$y - \text{coordinate, ie } f(x) = 2 \times \dfrac{9}{16} + 3 \times \dfrac{-3}{4} - 2 = -3\dfrac{1}{8}$$

Since 'a' = 2, the curve is concave and the point is a minimum point.

Cubic (3rd Degree) graph showing two turning points.

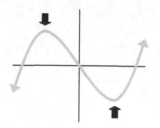

Cubic (3rd Degree) graph with no turning points

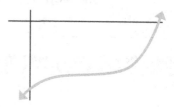

Quartic (4th Degree) graph with three turning points

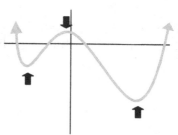

Quartic (4th Degree) graph with one turning point
The shapes of the graph of any polynomial function can be easily obtained using a graphics calculator.

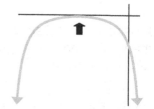

# INTERSECTION POINTS OF FUNCTIONS

When two lines meet each other we have a point of intersection. This point is common to both lines and will satisfy both equations of the lines. The coordinates of the point of intersection is what we found when we solved two simultaneous linear equations.

One or more points of intersection occur when a line crosses a curve. In the diagram below there are three points of intersection and the coordinates of these points can be obtained by solving the equation of the line with that of the curve.

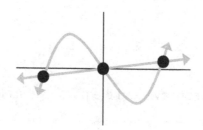

**Example:**

- Find the points of intersection of the two functions $y = x^3 + 3x^2 - 4x + 2$ and $y = 3x - 1$

Name the two functions follows:

$y = x^3 + 3x^2 - 4x + 2$ ------------------- (1)
$y = 3x - 1$ -------------------------------- (2)

Substituting this for y in equation (1) we get:

$3x - 1 = x^3 + 3x^2 - 4x + 2$

That is, $x^3 + 3x^2 - 7x + 3 = 0$

Using the factor theorem we see that $x - 1$ is a factor:

The other quadratic factor is found by long division as follows:

$$
\begin{array}{r}
x^2 \quad +4x \quad -3 \phantom{00000} \\
x-1 \overline{\smash{\big)}\, x^3 \ +3x^2 \ -7x \ +3} \\
\underline{x^3 \ -x^2 \phantom{000000000}} \\
4x^2 \ -7x \phantom{000} \\
\underline{4x^2 \ -4x \phantom{000}} \\
-3x \ +3 \\
\underline{-3x \ +3} \\
0
\end{array}
$$

Therefore, $(x - 1)(x^2 + 4x - 3) = 0$

We now solve $x^2 + 4x - 3 = 0$ using the quadratic formula.

$$x = \frac{-4 \pm \sqrt{16 + 12}}{2} = \frac{-4 \pm 5.291}{2}$$

$x = 0.65$ or $-4.65$

The solutions to the given equations are: $x = 1, 0.65$ and $-4.65$

The corresponding values of y are obtained by substituting in $y = 3x - 1$:

The y values are 2, 0.95 and $-14.95$. The coordinates of the points of intersection are:

$(1, 2), (0.65, 0.95)$ and $(-4.65, -14.95)$

The accuracy of the coordinates depends on the number of decimal places used. Two curves such as the parabolas can intersect. There can be either one common point or two common points in such intersections as shown below:

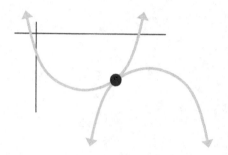 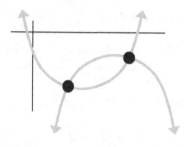

The point(s) of intersection can be found by solving the equations of the two curves.

**Examples:**

- Find the point of intersection of the two parabolas $y = x^2 - 4x - 5$ and $y = x^2 - 5x + 7$

Name the two functions as follows:

$y = x^2 - 4x - 5$ ------------------------(1)
$y = x^2 - 5x + 7$ ------------------------(2)

(1) – (2) gives: $0 = x - 12$

Therefore, $x = 12$

Substituting in (1) we get $y = 144 - 48 - 5 = 91$

Therefore the point of intersection is (12, 91)

- Find the point of intersection of the two parabolas $y = x^2 + 3x - 6$ and $y = -x^2 + 5x - 4$

Name the two functions as follows:

$y = x^2 + 3x - 6$ ------------------------(1)
$y = -x^2 + 5x - 4$ ------------------------(2)

(1) – 2 gives: $0 = 2x^2 - 2x - 2$

That is, $x^2 - x - 1 = 0$

134

$$x = \frac{+1 \pm \sqrt{1+4}}{2} = \frac{+1 \pm 2.236}{2}$$

x = 1.62 or –0.62

The corresponding values of y are 1.48 and –7.48

The coordinates of the points of intersection are: (1.62, 1.48) and (–0.62, –7.48)

## EXERCISE 27.1----------------------------------------------------------------

Using the remainder theorem or otherwise find the remainder for the following divisions:

1. $a^2 - 6a - 4 \div (a - 2)$

2. $6x^2 + 9x - 11 \div (x + 3)$

3. $p^3 - 11p + 5 \div (p + 2)$

4. $3m^3 - 4m - 13 \div (m - 3)$

5. $x^4 + 2x^3 - 3x^2 + 5x - 8 \div (x + 2)$

6. $2x^4 + 3x^3 - 4x^2 + 6x - 7 \div (x - 3)$

7. $a^4 + 3a^3 + 7a^2 + 6a + 7 \div (a^2 + 2a + 3)$

8. $y^4 + 3y^3 - 8y^2 + 7y - 4 \div (y^2 - 2y + 3)$

| Solutions: Exercise 27.1 | | | |
|---|---|---|---|
| **1.** –12 | **3.** 19 | **5.** –30 | **7.** –a + 1 |
| **2.** 16 | **4.** 56 | **6.** 218 | **8.** –10y – 1 |

## EXERCISE 27.2------------------------------------------------------------------

Using the factor theorem in conjunction with long division where required, find all the factors of each of the following:

1. $x^2 - 2x - 3$

2. $a^2 - 4a - 21$

3. $p^3 + 4p^2 + p - 6$

4. $m^3 + m^2 - 16m - 16$

5. $x^4 - 4x^3 + 9x^2 - 24x + 18$

6. $y^4 + y^3 - y - 1$

7. $x^3 - y^3$

8. $b^4 + 3b^3 - 3b^2 - 12b - 4$

| Solutions: Exercise 27.2 | | | |
|---|---|---|---|
| **1.** (x + 1)(x – 3) | **3.** (p–1)(p+2)(p+3) | **5.** (x–1)(x–3)($x^2$ + 6) | **7.** (x–y)($x^2$+xy+$y^2$) |
| **2.** (a + 3)(a – 7) | **4.** (m+1)(m+4)(m–4) | **6.** (y+1)(y–1)($y^2$+y+1) | **8.** (b+2)(b–2)($b^2$+3b+1) |

**EXERCISE 27.3**----------------------------------------------------------------

Determine the nature of the turning point of each of the following quadratic polynomial functions and calculate the maximum or minimum value of the function.

1.    $f(x) = x^2 - 3x - 4$

2.    $f(a) = -4a^2 - 8a + 5$

3.    $f(t) = 8t^2 - t - 3$

4.    $f(p) = -p^2 + p + 5$

5.    $f(m) = 6m^2 - 15m + 1$

6.    $f(x) = -x^2 - 3x + 8$

7.    $f(b) = b^2 - 4b - 10$

8.    $f(x) = -12x^2 + 2x + 7$

------------------------------------------------------------------------

| **Solutions: Exercise 27.3** | | | |
|---|---|---|---|
| **1.** Concave<br><br>Min. value $= -6\frac{1}{4}$ | **3.** Concave<br><br>Min. value $= -3\frac{1}{32}$ | **5.** Concave<br><br>Min. value $-8\frac{3}{8}$ | **7.** Concave<br><br>Min. value $-14$ |
| **2.** Convex<br><br>Max. value $= 9$ | **4.** Convex<br><br>Max. value $= 5\frac{1}{4}$ | **6.** Convex<br><br>Max. value $= 10\frac{1}{4}$ | **8.** Convex<br><br>Max. value $7\frac{1}{12}$ |

**EXERCISE 27.4**----------------------------------------------------------------

Determine the coordinates of the point(s) of intersection of the following pairs of functions giving answers corrected to two decimal places where necessary.

1.    $y = x + 1;\quad y = x^2$

2.    $y = 2x + 3;\quad y = x^2 + 4$

3.    $y = 3x - 2;\quad y = x^3$

4.    $y = 3x + 7;\quad y = 5x - 3$

5.    $y = x^2 + 4x;\quad y = 3x^2 - 6$

6.    $y = x^2 - 3x - 8;\quad y = x^2 + 4x + 6$

7.    $y = (x - 4)^2;\quad y = 3x^2 - 6$

8.    $y = 9 + 3x - 2x^2;\quad y = 27 - 12x - 4x^2$

------------------------------------------------------------------------

| **Solutions: Exercise 27.4** | | | |
|---|---|---|---|
| **1.** (1.62, 2.62)<br>(−0.62, 0.38) | **3.** (1, 1) twice<br>(− 2, − 8) | **5.** (3, 21)<br>(−1, − 3) | **7.** (1.87, 4.52)<br>(−5.87, 97.48) |
| **2.** (1, 5) twice | **4.** (5, 22) | **6.** (−2, 2) | **8.** (1.05, 9.94)<br>(−8.55, −162.81) |

# CHAPTER 28

A fraction can be expressed as the sum of multiple fractions:

$$\frac{3}{4} = \frac{1}{2} + \frac{1}{4} \quad \text{or} \quad \frac{3}{4} = \frac{1}{4} + \frac{1}{4} + \frac{1}{4}$$

Similarly, an algebraic fraction can also be split into the sum of two or more fractions. This lesson looks at some of the commonly encountered methods to express a fraction as partial fractions.

A fraction such as $\dfrac{x}{x+1}$ can be written as $\dfrac{(x+1)-1}{x+1}$. By dividing the $(x+1)$ and 1 in the numerator individually by $x+1$ in the denominator, the answer can be written as $1 - \dfrac{1}{x+1}$. The same result can be obtained by long division.

The general rule is to carry out long division if the highest power of the variable is equal to or greater in the numerator when compared to the denominator.

---

In general, $\dfrac{ax}{ax+1} = 1 - \dfrac{1}{ax+1}$

---

Let us consider the fraction $\dfrac{2x^2+3x-2}{x+1}$ where the highest power of the variable in the numerator is greater than that in the denominator. After carrying out long division as shown below we can write:

$$
\begin{array}{r}
2x \quad +1 \phantom{0} \\
x+1 \overline{\smash)\; 2x^2 \;+3x\; -2} \\
\underline{2x^2 \;+2x} \\
x \;-2 \\
\underline{x \;+1} \\
0
\end{array}
$$

Therefore, $\dfrac{2x^2+3x-2}{x+1} = (2x+1) - \dfrac{3}{x+1}$

The quadratic denominator in a fraction may or may not be factorised. If it can be factorised then we can easily express it in partial fractions.

**Example:**

- Express $\dfrac{2}{x^2 - 3x + 2}$ as partial fractions

$x^2 - 3x + 2$ can be factorised into $(x - 2)(x - 1)$.

The partial fractions equal to the given fraction are written as:

$$\frac{2}{x^2 - 3x + 2} = \frac{A}{x - 2} + \frac{B}{x - 1}$$

$$\frac{2}{(x - 2)(x - 1)} = \frac{A}{x - 2} + \frac{B}{x - 1}$$

Multiplying the equation throughout by $(x - 2)(x - 1)$, we get:

$2 = A(x - 1) + B(x - 2)$
$2 = Ax - A + Bx - 2B$
$2 = (A + B)x - (A + 2B)$

Equating the coefficient of x and the constant on both sides of the equation:

$A + B = 0$ -----------(1)
$-A - 2B = 2$ ---------(2)

Adding the above two equations we have:

$B = -2$ ------------(3)

Substituting (3) into (1) we get:

$A = -B = 2$

Therefore, $\dfrac{2}{x^2 - 3x + 2} = \dfrac{2}{x - 2} - \dfrac{2}{x - 1}$

The cover up method is a quick alternative for fractions that only have linear factors. We will now solve the previous example using the cover up method:

**Examples:**

- Express $\dfrac{2}{x^2 - 3x + 2}$ as partial fractions

$$\frac{2}{(x-2)(x-1)} = \frac{A}{x-2} + \frac{B}{x-1}$$

After writing the partial fractions as above, we first determine the value of A. This is done by equating $x - 2$ to zero and getting the value of x as 2. We now cover up $(x - 2)$ on the left hand side fraction and substitute $x = 2$ for the rest of the expression to determine the value of A

$$\frac{2}{2-1} = 2 = A$$

Now equate $x - 1$ to zero and get the value of x as 1. Cover up $(x - 1)$ on the left hand side fraction and substitute $x = 1$ for the rest of the expression and find the value of B.

$$\frac{2}{1-2} = -2 = B$$

Therefore, $\dfrac{2}{x^2 - 3x + 2} = \dfrac{2}{x-2} - \dfrac{2}{x-1}$

- Express $\dfrac{x}{x^2 - 1}$ as partial fractions

The denominator of the fraction can be factorised as $(x + 1)(x - 1)$

Therefore, $\dfrac{x}{(x+1)(x-1)} = \dfrac{A}{x+1} + \dfrac{B}{x-1}$

Multiplying the equation throughout by $(x + 1)(x - 1)$, we get:

$x = A(x - 1) + B(x + 1)$
$x = Ax - A + Bx + B$
$x = (A + B)x - (A - B)$

Equating the coefficient of x and the constant on both sides of the equation:

A + B = 1 --------(1)
−A + B = 0 ------(2)

Adding the above two equations we have:

B = 0.5 --------(3)

Substituting (3) into (2) we get:

A = B = 0.5

Therefore $\dfrac{x}{x^2-1} = \dfrac{0.5}{x+1} + \dfrac{0.5}{x-1} = \dfrac{1}{2(x+1)} + \dfrac{1}{2(x-1)}$

Note: Check that the cover up method gives the same values for A and B

## REPEATED FIRST DEGREE FACTOR IN THE DENOMINATOR

The partial fractions are written as follows when a factor is repeated in the denominator of the original fraction.

**Examples:**

- Express $\dfrac{3x^2+5x-6}{(x+3)(x-1)^2}$ as partial fractions

There are two (x − 1) factors in the denominator of the fraction so we write

$$\dfrac{3x^2+5x-6}{(x+3)(x-1)^2} = \dfrac{A}{(x+3)} + \dfrac{B}{(x-1)} + \dfrac{C}{(x-1)^2}$$

To determine the values of A, B and C we proceed as before and equate the coefficient of $x^2$ and x and the constants on either side. Note that the cover up method cannot be used here.

- Express $\dfrac{2x^2-5x+3}{(x-1)(x+2)^3}$ as partial fractions

There are three (x + 2) factors in the denominator of the fraction so we write

$$\dfrac{2x^2-5x+3}{(x-1)(x+2)^3} = \dfrac{A}{(x-1)} + \dfrac{B}{(x+2)} + \dfrac{C}{(x+2)^2} + \dfrac{D}{(x+2)^3}$$

When the denominator of the fraction contains both a linear and quadratic factor we write the partial fractions as shown below:

**Examples:**

- Express $\dfrac{2x+7}{x(x^2+5x+7)}$ as partial fractions

$$\frac{2x+7}{x(x^2+5x+7)} = \frac{A}{x} + \frac{Bx+C}{(x^2+5x+7)}$$

Note that the partial fraction with the quadratic denominator has a linear factor in the numerator. To find the values of A, B and C we multiply the equation throughout by $x\,(x^2+5x+7)$ and get:

$2x + 7 = A(x^2 + 5x + 7) + x(Bx + C)$
$2x + 7 = Ax^2 + 5Ax + 7A + Bx^2 + Cx$
$2x + 7 = (A + B)x^2 + (5A + C)x + 7A$

Equating the coefficients of $x^2$ and $x$ and constants:

$A + B = 0$
$5A + C = 2$
$7A = 7$

Solving the equations we get A = 1, B = –1 and C = –3.

Hence, $\dfrac{2x+7}{x(x^2+5x+7)} = \dfrac{1}{x} - \dfrac{x+3}{(x^2+5x+7)}$

- Express $\dfrac{3a^2-2a+1}{(a^2+1)(a^2+2)}$ as partial fractions

$$\frac{3a^2-2a+1}{(a^2+1)(a^2+2)} = \frac{Aa+B}{(a^2+1)} + \frac{Ca+D}{(a^2+2)}$$

Multiplying the equation throughout by $(a^2 + 1)\,(a^2 + 2)$, we get:

$3a^2 – 2a + 1 = (a^2 + 2)(Aa + B) +(a^2 + 1)(Ca + D)$
$3a^2 – 2a + 1 = Aa^3 + Ba^2 + 2Aa + 2B + Ca^3 + Da^2 + Ca + D$
$3a^2 – 2a + 1 = (A + C)a^3 +(B + D)a^2 +(2A + C)a +(2B + D)$

Equating the coefficients of $a^3$, $a^2$, $a$ and the constants, we get

$A + C = 0$ --------------(1)
$B + D = 3$ --------------(2)
$2A + C = -2$ -------------(3)
$2B + D = 1$ -------------(4)

Solving the equations we get $A = -2$, $B = -2$, $C = 2$, $D = 5$.

Hence, $\dfrac{3a^2 - 2a + 1}{(a^2 + 1)(a^2 + 2)} = \dfrac{-2a - 2}{(a^2 + 1)} + \dfrac{2a + 5}{(a^2 + 2)} = \dfrac{-2(a + 1)}{(a^2 + 1)} + \dfrac{2a + 5}{(a^2 + 2)}$

## EXERCISE 28.1

Express the following as partial fractions:

1. $\dfrac{x}{x + 6}$

3. $\dfrac{5}{(x + 2)(x - 1)}$

5. $\dfrac{9m - 2}{2m^2 - 7m + 3}$

7. $\dfrac{4}{c^2 + c}$

2. $\dfrac{4a}{4a + 3}$

4. $\dfrac{7a}{(a + 3)(a - 2)}$

6. $\dfrac{3p - 2}{p^2 - 4}$

8. $\dfrac{2x - 1}{(x - 3)(x + 1)}$

| Solutions: Exercise 28.1 | |
|---|---|
| 1. $1 - \dfrac{6}{x + 6}$ | 5. $\dfrac{-1}{2m - 1} + \dfrac{5}{m - 3}$ |
| 2. $1 - \dfrac{3}{4a + 3}$ | 6. $\dfrac{2}{p + 2} + \dfrac{1}{p - 2}$ |
| 3. $\dfrac{-5}{3(x + 2)} + \dfrac{5}{3(x - 1)}$ | 7. $\dfrac{4}{c} - \dfrac{4}{(c + 1)}$ |
| 4. $\dfrac{21}{5(a + 3)} + \dfrac{14}{5(a - 2)}$ | 8. $\dfrac{5}{4(x - 3)} + \dfrac{3}{4(x + 1)}$ |

## EXERCISE 28.2

Express the following as partial fractions:

1. $\dfrac{11y + 7}{y^2 - 1}$

3. $\dfrac{5a - 21}{(a - 5)^2}$

5. $\dfrac{6 - x}{(1 - x)(3 - x)}$

7. $\dfrac{p^2 + 2p + 4}{p^2 - 4p}$

2. $\dfrac{4a - 7}{(a - 2)^2}$

4. $\dfrac{x^3 + 3x^2 + 5x + 3}{x^2 - 5x + 6}$

6. $\dfrac{8}{x^3 + 2x}$

8. $\dfrac{2b^2 - 9b + 7}{b^2 - 6b + 9}$

## Solutions: Exercise 28.2

| | |
|---|---|
| **1.** $\dfrac{2}{y+1} + \dfrac{9}{y-1}$ | **5.** $\dfrac{5}{2(1-x)} - \dfrac{3}{2(3-x)}$ |
| **2.** $\dfrac{4}{a-2} + \dfrac{1}{(a-2)^2}$ | **6.** $\dfrac{4}{x} - \dfrac{4x}{x^2+2}$ |
| **3.** $\dfrac{5}{a-5} + \dfrac{4}{(a-5)^2}$ | **7.** $1 - \dfrac{1}{p} + \dfrac{7}{p-4}$ |
| **4.** $x+8 - \dfrac{33}{x-2} + \dfrac{72}{x-3}$ | **8.** $2 + \dfrac{3}{b-3} - \dfrac{2}{(b-3)^2}$ |

## EXERCISE 28.3---------------------------------------------------------------------

Decompose each of the following into partial fractions:

**1.** $\dfrac{3m+1}{(m-1)(2m+1)}$

**2.** $\dfrac{3x+1}{(x-1)^2(x+2)}$

**3.** $\dfrac{7a^2-1}{a+3}$

**4.** $\dfrac{c^2-2}{c(c^2+1)}$

**5.** $\dfrac{5p+11}{p^2+4p+3}$

**6.** $\dfrac{5b}{(b^2+b+1)(b-2)}$

**7.** $\dfrac{y}{(y^2-y+1)(3y-2)}$

**8.** $\dfrac{3d^2+4d+7}{9(d-1)^2}$

## Solutions: Exercise 28.3

| | |
|---|---|
| **1.** $\dfrac{4}{3(m-1)} + \dfrac{1}{3(2m+1)}$ | **5.** $\dfrac{3}{p+1} + \dfrac{2}{p+3}$ |
| **2.** $\dfrac{5}{9(x-1)} + \dfrac{4}{3(x-1)^2} - \dfrac{5}{9(x+2)}$ | **6.** $\dfrac{-10b+5}{7(b^2+b+1)} + \dfrac{10}{7(b-2)}$ |
| **3.** $7a-21 + \dfrac{62}{a+3}$ | **7.** $\dfrac{-2y+3}{7(y^2-y+1)} + \dfrac{6}{7(3y-2)}$ |
| **4.** $\dfrac{-2}{c} + \dfrac{3c}{c^2+1}$ | **8.** $\dfrac{1}{3} + \dfrac{10}{9(d-1)} + \dfrac{14}{9(d-1)^2}$ |

# CHAPTER 29

## PATTERNS, SEQUENCES AND SERIES

### PATTERNS

A **pattern** is a general term for any recognisable regularity in a given observation. Examples where patterns occur include weather, designs and number sequences. All patterns follow some order or rules. There are linear patterns and non-linear patterns. A pattern can be represented in a number of ways ie multiple representations.

### PATTERNS THAT CONTINUE INDEFINITELY

A pattern that keeps on recurring is called a **repeating pattern.**

**Examples:**

- R G Y   R G Y   R G Y   R G Y ............
- White, Black, Red, White, Black, Red ..........
- 1, 2, 3, 4, 1, 2, 3, 4, 1, 2, 3, 4 .............

### PATTERNS THAT GROW

When one or more terms or elements in a pattern build in a systematic way to form a larger pattern, the result is a **growth pattern.**

- RGY, RRGGYY, RRRGGGYYY .............

- 01, 0011, 000111, 00001111..........

- ▲ , ▲▲ , ▲▲▲ ..........

### NUMBER PATTERNS

**Number patterns** are generally limited to growth patterns where the numerical value of each element is important.

For example, 2, 5, 8, 11, 14 ........ is a number pattern where each element is 3 more than the preceding element.

In order to find a missing number in a sequence we do the following:

- Determine if the order of numbers is ascending (getting larger in value) or descending (becoming smaller in value).
- Find the difference between numbers that are next to each other.
- Use the difference between numbers to find the missing number.

**Example:**

Find the missing number: 15, 13, x, 9

The difference between the first two numbers is 2. Therefore x is 11.

## POSITION NUMBERS

Consider the following pattern and the position numbers of each of the three elements A, B and C:

| Element: | A | B | C | A | B | C | A | B | C | A | B | C | A | B | C |
|---|---|---|---|---|---|---|---|---|---|---|---|---|---|---|---|
| Position Number | 1 | 2 | 3 | 4 | 5 | 6 | 7 | 8 | 9 | 10 | 11 | 12 | 13 | 14 | 15 |

A has positions 1, 4, 7, 10 ……….

B has positions 2, 5, 8, 11 ……….

C has positions 3, 6, 9, 12………..

Each element's position number increases by 3 with reference to its previous position.

The same pattern can be presented in different ways. Whole numbers can be presented in various forms such as a **numerical sequence**, **graph**, or **chart.**

## DESCRIBING PATTERNS OR CHANGE BY A SIMPLE RULE

In the following sets of number patterns, the words in brackets describe each pattern. This is called the **rule**.

**Examples:**

- 1, 6, 11, 16, ___, ___    (add 5)
- 2, 8, 2, 8, ___, ___    (alternating numbers)
- 28, 24, 20, 16, ___, ___    (subtract 4)
- 5, 10, 15, 20, ___, ___ (skip count by 5)
- 3, 5, 8, 13, 21, 34, ___, ___ (add the two previous numbers)

The last pattern is a **recursive pattern** known as the **Fibonacci sequence.**

**More examples:**

- $1 \longrightarrow 3 \longrightarrow 7 \longrightarrow$   (multiply by 2 and then add 1)

- $1 \longrightarrow 4 \longrightarrow 10 \longrightarrow$   (multiply by 2 and then add 2)

- $3 \longrightarrow 4 \longrightarrow 6 \longrightarrow$   (subtract 1 and then multiply by 2)

- $4 \longrightarrow 5 \longrightarrow 7 \longrightarrow$   (multiply by 2 and then subtract 3)

- $8 \longrightarrow 10 \longrightarrow 14 \longrightarrow$   (subtract 3 and then multiply by 2)

**EXERCISE 29.1**---------------------------------------------------------------------------------

1. John puts three different types of symbols on a straight line as shown below. Which two symbols continue this pattern?

ÐÑÐÇÐÑÐÇÐÑÐÇ??

2. Susan laid the following figures on the floor by repeating the patterns shown below covering a length of 52cm. How many squares did she use?

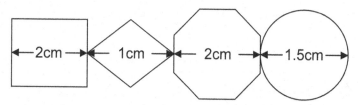

3. Study the following pattern and answer the question below:

$X_2 = 2 + 1$
$X_3 = 3 + 2 + 1$
$X_4 = 4 + 3 + 2 + 1$

   What is the value of $X_6 - X_2$?

4. Complete the following pattern:

   $3 \longrightarrow 5; \quad 4 \longrightarrow 10; \quad 5 \longrightarrow 15 \ldots\ldots\ldots\ldots; \quad 10 \longrightarrow ?$

5. The size of trousers and waist measurements follow a pattern as shown below.

| Size | 6 | 7 | 8 | 9 | 10 | 11 | 12 |
|------|------|------|------|------|------|------|------|
| Waist | 61.5 | 64 | 66.5 | 69 | ? | ? | ? |

   What is the waist measurement for size 12?

6.  What is the next number in the following number pattern?

2, 3, 5, 8, 12, ?

7.  What is the next term in the number pattern 91, 45, 22, ...........?

8.  Fill in the missing figures in the following table:

| 4 | 5 | 9 | ? | ? | ? | ? | ? | ? |
|---|---|---|---|---|---|---|---|---|
| 2 | 2 | 4 | 7 | 5 | 7 | 7 | 1 | 3 |
| 2 | 3 | 5 | 6 | 5 | 8 | 9 | 2 | 4 |
| 4 | 6 | ? | ? | ? | ? | ? | ? | ? |

------------------------------------------------------------------------------

| Solutions: Exercise 29.1 | |
|---|---|
| **1.** Đ and Ñ | **5.** 76.5 |
| **2.** 8 | **6.** 17 |
| **3.** 18 | **7.** 10.5 |
| **4.** 6-20; 7-25; 8-30; 9-35; 10-40 | **8.** Top row: 13,10, 15, 16, 3, 7 |
| | Bottom row: 20, 42, 25, 56, 63, 2, 12 |

## SEQUENCES

Sequences are a string of numbers that follow a set of rules which determine the number (or term) next to each other. Any particular term in the sequence is referred to as the $n^{th}$ term. It is usually represented by the notation $a_n$ or $T_n$. For example, the $3^{rd}$ term is represented by $a_3$ or $T_3$.

- **Arithmetic Sequence:** This is a sequence where a term is obtained by adding a fixed number to the immediately preceding term.

The pattern 3, 5, 7, 9, 11, 13….is an arithmetic sequence where the fixed number is 2. The fixed number added is called the 'common difference".

- **Geometric sequence:** This is a sequence where a term is obtained by multiplying the immediately preceding term by a fixed number.

The pattern 2, 6, 18, 54, 162…is a geometric sequence where the fixed number is 3. The fixed number that multiplies is called the "common ratio".

- **Fibonacci sequence:** This is a sequence where two immediately preceding terms are added to get the next term.

The pattern 1, 1, 2, 3, 5, 8, 13 ……is referred to as a Fibonacci sequence.

When the sequence is from a defined set of numbers we refer to it as a **finite sequence.** If the set is open (eg all positive integers) then a sequence from that set is an **infinite sequence.**

## FINDING THE SEQUENCE WHEN ITS N^TH TERM IS KNOWN

Suppose we have the formula for the nth term as $T_n = 2n - 1$. Let us substitute 1, 2 and 3, in turn for n:

$T_1 = 2(1) - 1 = 1$
$T_2 = 2(2) - 1 = 3$
$T_3 = 2(3) - 1 = 5$

The sequence for which the nth term is $2n - 1$ is: 1, 3, 5 .........

## FINDING THE N^TH TERM IN A SEQUENCE

**Examples:**

- Find the $n^{th}$ term of the sequence 4, 7, 10, 13 .........

We notice that any term in the sequence exceeds the preceding term by 3. Now we take the 4 which is term 1 and multiply 1 by 3 to get 3. We need to add one to this result to get the second term 4. We check to see whether this formula works for the next term as well:

$3 \times 2 + 1 = 7$

The formula works, so we can see that $3n + 1$ is the $n^{th}$ term. This is expressed as:

$T_n = 3n + 1$

- What term would 58 be in the sequence 4, 7, 10, 13 ...

We can use the formula for the nth term to answer this question.

$T_n = 3n + 1$
$58 = 3n + 1$
$3n = 57$
$n = 19$

Therefore, the $19^{th}$ term in the given sequence is 58.

- Find the nth term of the sequence 7, 11, 15, 19 ............

The difference between consecutive terms is 4.

Consider the 4 times table 4, 8, 12, 16, 20 ...which can be stated as 4n where n = 1, 2, 3 etc. Add 3 to 4n to find the nth term of the sequence. Therefore, the $n^{th}$ term of the given sequence is $T_n = 4n + 3$

# METHOD OF FINITE DIFFERENCES

The nth term of a sequence can be determined by calculating the differences between successive terms and then the differences between successive differences themselves. We stop when constant differences are obtained.

The relationship will be linear (ie of the form $T_n = an + b$) if the first differences are constant.

The relationship will be quadratic (i.e of the form $T_n = an^2 + bn + c$) if the second differences are constant.

The relationship will be cubic (i.e of the form $an^3 + bn^2 + cn + d$) if the third differences are constant.

**Examples:**

- Consider the number sequence: $-1, 1, 3, 5, 7, 9$ ............

The differences between successive terms are: 2, 2, 2, 2, 2

We have obtained constant difference of 2 between successive terms. Therefore, the relationship is $T_n = an + b$.

We need to determine the values of a, b. This is done by substituting any two numbers from the given sequence. If we choose the first two numbers, we have the following equations;

$-1 = a + b$ -------------------(1)
$1 = 2a + b$ -------------------(2)

(2) – (1) gives $2 = a$ ---------------(3)

Substituting (3) into (1), we get $b = -3$

Therefore, $T_n = 2n - 3$

- Consider the number sequence: $5, 12, 21, 32, 45, 60$ ............

The **first differences** between successive terms are 7, 9, 11, 13, 15
The **second differences** between successive differences are 2, 2, 2, 2

We have obtained a constant difference of 2 between successive differences. Therefore, the relationship is $T_n = an^2 + bn + c$

We need to determine the values of a, b and c. This is done by substituting any three numbers from the given sequence. If we choose the first three numbers, we have three equations.

$5 = a + b + c$ ----------------------(1)
$12 = 4a + 2b + c$ -------------------(2)
$21 = 9a + 3n + c$ ------------------(3)

$(2) - (1)$ gives $7 = 3a + b$ -----------------------(4)
$(3) - (2)$ gives $9 = 5a + b$ -----------------------(5)
$(5) - (4)$ gives $2 = 2a$

Therefore, $a = 1$

Substituting $a = 1$ in (4) we get $b = 4$

Substituting for a and b in (1) we get $c = 0$

Therefore, $T_n = n^2 + 4n$

- Consider the number sequence: 4, 11, 30, 67, 128, 219…… …………

The **first differences** between successive terms are 7, 19, 37, 61, 91
The **second differences** between successive differences are 12, 18, 24, 30
The **third differences** between successive differences are 6, 6, 6

We have obtained constant difference of 6 between successive terms. Therefore, the relationship is $T_n = an^3 + bn^2 + cn + d$. We need to determine the values of a, b, c and d. This is done by substituting any four numbers. If we choose the first four numbers, we have the following equations:

$4 = a + b + c + d$ ----------------------(1)
$11 = 8a + 4b + 2c + d$ ---------------(2)
$30 = 27a + 9b + 3c + d$ -------------(3)
$67 = 64a + 16b + 4c + d$ -----------(4)

$(2) - (1)$ gives $7 = 7a + 3b + c$ --------------------(5)
$(3) - (2)$ gives $19 = 19a + 5b + c$ --------------------(6)
$(4) - (3)$ gives $37 = 37a + 7b + c$ --------------------(7)
$(6) - (5)$ gives $12 = 12a + 2b$ ----------------------(8)
$(7) - (6)$ gives $18 = 18a + 2b$ ----------------------(9)

$(9) - (8)$ gives $6 = 6a$

Therefore, $a = 1$

Substituting for 'a' in (8) we get $b = 0$

Substituting in (5) for 'a'; and 'b' we get $c = 0$

Substituting in (1) for 'a', 'b' and 'c' we get $d = 3$;

Therefore, $T_n = n^3 + 3$

# THE FACTORIAL NOTATION

The product of consecutive numbers from n to 1 can be written:

n(n − 1)(n − 2)(n − 3)(n − 4)………..1

A simpler way to indicate the above product is to write n! ('n' followed by an exclamation mark). It is pronounced 'n' factorial.

**Examples:**

- $2! = 2(1) = 2$

- $5! = (5)(4)(3)(2)(1) = 120$

- $6! = 6(5)(4)(3)(2)(1) = 720$

# THE $^N C_R$ NOTATION

This notation $^n c_r$ stands for $\dfrac{n!}{r!.(n-r)!}$.

**Example:**

$$^5c_3 = \frac{5!}{3!(2!)} = \frac{5(4)(3)(2)(1)}{3(2)(1).(2)(1)} = 10 \ (n = 5 \text{ and } r = 3)$$

The meaning of the answer 10 can be interpreted as the possible number of combinations of 3 things taken at a time that can be made from 5 things. If we apply this meaning to calculate the possible number of combination of 5 things taken at a time from 5 things, the answer would 1.

Therefore, $^5c_5 = 1$. That is, $\dfrac{5!}{5!(0!)} = 1$. This means that $0! = 1$

The expansion of $^5c_3$ by definition is:

$$\frac{5!}{(5-3)! \times 3!} = \frac{5 \times 4 \times 3 \times 2 \times 1}{(2 \times 1)(3 \times 2 \times 1)} = \frac{5 \times 4 \times 3}{(3 \times 2 \times 1)} = \frac{5 \times 4}{(2 \times 1)} = 10$$

The short cut to calculating $^5c_3$ is to multiply 3 consecutive numbers starting from 5 and divide the result by 3!. $^5c_3$ is also the same as $^5c_{5-3}$ which is equal to $^5c_2$.

**More examples:**

- $^6C_4 = {}^6C_2 = \dfrac{6 \times 5}{2 \times 1} = 15$

- $^{10}C_7 = {}^{10}C_3 = \dfrac{10 \times 9 \times 8}{3 \times 2 \times 1} = 120$

- $^{20}C_{18} = {}^{20}C_2 = \dfrac{20 \times 19}{2 \times 1} = 190$

## EXERCISE 29.2------------------------------------------------------------

Find the nth term of the following sequences:

1. 0, 1, 3, 6, 10 …

2. –6, –5, –2, 3, 10 …

3. 5, 6, 7, 8, 9 ….

4. 7, 9, 15, 25, 39 …

5. 2, 0, 0, 2, 6 …

6. –1, 2, 9, 20, 35 …

7. –1, 0, 7, 26, 63…

8. 0, 2, 9, 30, 74…

--------------------------------------------------------------------------

| Solutions: Exercise 29.2 | |
|---|---|
| **1.** $\dfrac{1}{2}n(n-1)$ | **5.** $n^2 - 5n + 6$ |
| **2.** $n^2 - 2n - 5$ | **6.** $n(2n - 3)$ |
| **3.** $n + 4$ | **7.** $n^3 - 3n^2 + 3n - 2$ |
| **4.** $2n^2 - 4n + 9$ | **8.** $\dfrac{1}{2}(3n^3 - 13n^2 + 22n - 12)$ |

## EXERCISE 29.3------------------------------------------------------------

Evaluate the following:

1. $5!$

2. $8!$

3. $\dfrac{10!}{6!}$

4. $\dfrac{7!}{2!4!}$

5. $\dfrac{8!}{4!(8-4)!}$

6. $\dfrac{9!}{0!(9-0)!}$

7. $\dfrac{14!}{11!3!}$

8. $\dfrac{22!}{20!2!}$

--------------------------------------------------------------------------

| Solutions: Exercise 29.3 | |
|---|---|
| **1.** 120 | **5.** 70 |
| **2.** 40320 | **6.** 1 |
| **3.** 5040 | **7.** 364 |
| **4.** 105 | **8.** 231 |

**EXERCISE 29.4**----------------------------------------------------------------------------

Evaluate the following:

1. $^5c_2$     3. $^9c_4$     5. $^8c_6$     7. $^{11}c_{10}$

2. $^{10}c_8$     4. $^{22}c_{20}$     6. $^{15}c_{13}$     8. $^7c_0$

----------------------------------------------------------------------------

| Solutions: Exercise 29.4 | |
|---|---|
| 1.  10 | 5.  28 |
| 2.  45 | 6.  105 |
| 3.  126 | 7  11 |
| 4.  231 | 8.  1 |

## SERIES

A **series** is the sum of a finite or infinite set of numbers in a sequence.

The Greek alphabet capital sigma is used to represent the sum of a sequence. This is known as **the sigma notation**.

- $\sum_{i=1}^{4} 3r$ means the sum of 3, 6, 9 and 12 (ie substitute 1 to 4 for r in 3r)

- $\sum_{i=1}^{100} i$ represents the sum of a finite sequence of numbers from 1 to 100.

- $\sum_{1}^{\infty} i$ represents the sum of an infinite sequence of numbers.

**Example:**

Write in detail the sum expressed by $\sum_{1}^{6} r^2$

$$\sum_{1}^{6} r^2 = 1^1 + 2^2 + 3^2 + 4^2 + 5^2 + 6^2$$

## ARITHMETIC SERIES OR PROGRESSION (AP)

We have seen that an arithmetic sequence has a common difference. Let us call the first term of an arithmetic sequence to be 'a' and the common difference to be 'd'.

The sum ($S_n$) of such a sequence with n terms and the last term 'l' can be written as:

$$S_n = a + (a + d) + (a + 2d) \ldots\ldots\ldots (l - 2d) + (l - d) + l \text{ --------(1)}$$

We can write the above starting with 'l' as follows:

$$S_n = l + (l - d) + (l - 2d) \ldots\ldots\ldots (a + 2d) + (a + d) + a \text{ ---------(2)}$$

Adding (1) and (2):

$$2S_n = (a + l) + (a + l) + (a + l) \ldots\ldots\ldots\ldots\ldots(a + l) + (a + l) + (a + l)$$

There are n such (a + l) and therefore,

$$2S_n = n(a + l)$$

Hence, $S_n = \dfrac{n}{2}(a + l)$

In other words, the sum of an AP is equal to half the number of terms multiplied by the sum of the first term and the last term.

We know that 'l' is the $n^{th}$ term $T_n$ and 'l' can be formulated as $T_n = a + (n - 1)d$ as when n is replaced by 1, 2, 3 etc we obtain the various terms of the sequence. Thus we can state the formula for the sum of AP, $S_n$ as:

$$S_n = \dfrac{n}{2}\{a + a + (n - 1)\, d\} \text{ or } S_n = \dfrac{n}{2}\{2a + (n - 1)\, d\}$$

The formula for the common difference d can be stated as $d = T_n - T_{n-1}$.

**Examples:**

- What is the $8^{th}$ term of the Arithmetic Series: 3 + 5 + 7 + ……….

Here, a = 3 and d = 2    (d = 5 – 3 or 7 – 5)

$T_n = a + (n - 1)d$
$T_8 = 3 + (8 - 1) \times 2$
$T_8 = 3 + 14 = 17$

- Find the first term and the common difference of an Arithmetic Progression if the $6^{th}$ term is –10 and the $10^{th}$ term is –22.

$T_n = a + (n – 1) d$
$–10 = a + (6 – 1) \times d$
$–10 = a + 5d$ ------------------(1)
Similarly, $–22 = a + 9d$ ----------------(2)

(1) – (2) gives: $12 = –4d$

$d = –3$. Substituting 'd' into (1) we get

$–10 = a + 5 \times –3$
$a = 5$

Therefore the first term of the AP is 5 and the common difference is –3

- Find the sum of $1 + 2 + 3 + 4 + 5 + \ldots\ldots + 80$

$S_n = \dfrac{n}{2} (a + l)$

$S_n = \dfrac{80}{2} (1 + 80)$

$S_n = 40 \times 81$
$S_n = 3240$

- Find the sum of $4 + 7 + 10 + \ldots\ldots\ldots$ (50th term)

Here, $n = 50$, $a = 4$ and $d = 3$
$S_n = \dfrac{n}{2} \{2a + (n – 1)d\}$

$S_n = \dfrac{50}{2} \{2 \times 4 + (50 – 1) \times 3\}$

$S_n = 25(8 + 49 \times 3)$
$S_n = 25 \times 155$
$S_n = 3875$

The **arithmetic mean** of any two numbers is a number between the two numbers that will form an Arithmetic Progression.

---

**Example:**

- Find the arithmetic mean of 8 and 24

Let the arithmetic mean be x.  Then 8, x and 24 will form an AP.

That is, the common difference between the first two numbers and the last two numbers are equal.  We can express this as the following:

$x - 8 = 24 - x$
$2x = 24 + 8 = 32$
$x = 16$

The arithmetic mean = 16.  We have thus inserted 16 between 8 and 24 so that all three numbers for an AP.

---

The Arithmetic mean of two numbers A and B = $\frac{1}{2}(A + B)$

---

We can also insert more than one arithmetic mean between any two numbers so that all the numbers together form an AP.

---

**Example:**

- Insert three arithmetic means between 5 and 21

Let the three arithmetic means be $x_1$, $x_2$ and $x_3$

Therefore 5, $x_1$, $x_2$, $x_3$, 21 form an AP.  That is, a = 5 and $T_5$ = 21.

$T_5 = a + 4d$
$21 = 5 + 4 \times d$
$4d = 16$
$d = 4$

Therefore, $x_1 = 9$, $x_2 = 13$, and $x_3 = 17$

The three Arithmetic Means are 9, 13 and 17

---

Find the tenth term and the sum of the first ten terms for each of the following Arithmetic Progressions:

1. 4, 7, 10, 13 …

2. 3, 7, 11, 15 …

3. 25, 20, 15, 10….

4. −6, −9, −12. −15…

5. 3, 8, 13, 18…

6. $\frac{1}{2}$, 1, $1\frac{1}{2}$, 2…

7. 1, $\frac{5}{3}$, $\frac{7}{3}$, 3 …

8. 4, $3\frac{3}{4}$, $3\frac{1}{2}$, $3\frac{1}{4}$ …

| Solutions: Exercise 29.5 | |
|---|---|
| **1.** 31, 175 | **5.** 48, 255 |
| **2.** 39, 210 | **6.** 5, 27.5 |
| **3.** −20, 25 | **7.** 7, 40 |
| **4** −33, −195 | **8.** $1\frac{3}{4}$, $28\frac{3}{4}$ |

**EXERCISE 29.6**----------------------------------------------------------------------------------

Using the formula $T_n = a + (n − 1)d$, find the number of terms in each of the following Arithmetic Series. Find also the sum in each case using the formula $S_n = \frac{n}{2}(a + l)$

1. 1 + 2 + 3 +……… + 10

2. 5 + 6 + 7 + ……… + 55

3. 2 + 5 + 8 + ……… + 41

4. 3 + 5 + 7 + ………. + 53

5. 7 + 2 − 3 − 8 − ………. −68

6. 5 + 6.1 + 7.2 +……….. + 23.7

7. 3 + 1.95 + 0.9………. −7.5

8. $3\frac{1}{2}$ + 4 + $4\frac{1}{2}$ + …………. + 15

| Solutions: Exercise 29.6 | |
|---|---|
| **1.** 10, 55 | **5.** 16, −488 |
| **2.** 51, 1530 | **6.** 18, 258.3 |
| **3.** 14, 301 | **7.** 11, −24.75 |
| **4.** 26, 728 | **8.** 24, 222 |

**EXERCISE 29.7**----------------------------------------------------------------------

Use the formula $S_n = \dfrac{n}{2}\{2a + (n-1)d\}$ to find the sum of each of the following AP's:

1. $2 + 7 + 12 + \ldots$to 10 terms

5. $-8 - 11 - 14 - \ldots$to 9 terms

2. $6 + 8 + 10 + \ldots$to 12 terms

6. $1\dfrac{1}{6} + 2 + 2\dfrac{5}{6} + \ldots$ to 8 terms

3. $3 + 5 + 7 + \ldots$to 20 terms

7. $4 + 6.5 + 9 + \ldots$to 18 terms

4. $8 + 9\dfrac{1}{2} + 11 + \ldots$to 14 terms

8. $-4.4 - 2.3 - 0.2 + \ldots$to 16 terms

--------------------------------------------------------------------------

| Solutions: Exercise 29.7 | |
|---|---|
| **1.** 245 | **5.** $-180$ |
| **2.** 204 | **6.** $32\dfrac{2}{3}$ |
| **3.** 440 | **7.** 454.5 |
| **4.** $248\dfrac{1}{2}$ | **8.** 181.6 |

**EXERCISE 29.8**----------------------------------------------------------------------

Insert the indicated number of Arithmetic Means between each of the following pairs of numbers:

1. 8 and 20   (1 mean)

5. 4 and 54   (9 means)

2. $-3$ and 15   (1 mean)

6. 6.333 and 11   (6 means)

3. 3 and 39   (5 means)

7. 7.8 and 21   (5 means)

4. 12 and 44   (7 means)

8. $-5.65$ and $-17.8$   (8 means)

-------------------------------------------------------------------------

| Solutions: Exercise 29.8 | |
|---|---|
| **1.** 14 | **5.** 9, 14, 19, 24, 29, 34, 39, 44, 49 |
| **2.** 6 | **6.** $7, 7\dfrac{2}{3}, 8\dfrac{1}{3}, 9, 9\dfrac{2}{3}, 10\dfrac{1}{3}$ |
| **3.** 9, 15, 21, 27, 33 | **7** 10, 12.2, 14.4, 16.6, 18.8 |
| **4.** 16, 20, 24, 28, 32, 36, 40 | **8.** $-7, -8.35, -9.7, -11.05, -12.4, -13.75,$ $-15.1, -16.45$ |

**EXERCISE 29.9**----------------------------------------------------------------------

1. In an arithmetic progression with a common difference of 3, the 12$^{th}$ term is 38. Find the first term.

2. In an arithmetic progression, the fourth term is 21 and the 9$^{th}$ term is 56. Find the 15$^{th}$ term.

3. In an arithmetic progress the first term is –7 and the 14th term is 58. Find the sum of the first 12 terms.

4. The sum of n terms of an arithmetic progression is $\frac{n}{2}(3n - 15)$. What are the first four terms?

5. The ninth term of an Arithmetic progression is 1 and the sum of the first five terms equals the sum of the first eleven terms. Find the 17$^{th}$ term.

6. How many terms of the series 3 + 8 + 13 + 18 + …… will add up to 255?

7. Find the sum of all the integers between 1 and 100 that are divisible by 7.

8. The sum of n terms of an AP is $\frac{n}{2}(3n + 7)$. What is the 8$^{th}$ term?

-----------------------------------------------------------------------------------

| Solutions: Exercise 29.9 | |
|---|---|
| **1.** a = 5 | **5.** T$_{17}$ =17 |
| **2.** T$_{15}$ = 98 | **6.** 10 |
| **3.** 246 | **7.** 735 |
| **4.** – 6, – 3, 0, 3 | **8.** 26 |

**EXERCISE 29.10**---------------------------------------------------------------------

1. Peter deposited $100 in his bank savings account on the first day of January and thereafter made deposits every month increasing the previous month's deposit by $40 of $140, $180, $220 and so on. How many such monthly deposits are required to be made to have a total deposit of $5700 in the account.

2. Gillian was hired by a company at a salary of $80000 per year, with annual increases of $1,800, what will she earn in the sixth year?

3, A company's sales were $120000 in the first year of its operations but increased by $20000 each year thereafter. Calculate what the company's sales would be in its 8th year. What would be the total sales over the 8 years?

4. A man repaid a debt by annual instalments with each instalment increasing by $25 of the previous one. If the first and the last instalment were $1000 and $1225 respectively, how many annual instalments did he make? What was the total amount of the debt?

5. The population of a town increased each decade in an arithmetic progression. The sum of the population at the end of each decade for three consecutive decades is 3 million. The sum of the squares of the population at the end of each of the three decades is 3.5million. What was the population at the end of each of the three decades?

6. A machine purchased at the beginning of the year depreciates each year by $2000. If the depreciated value of the machine after 8 years is three-fifth of the original cost, what was its original cost?

7. The total weight of 12 packages is 4.98kg. The weights of consecutive packages differ by 0.03kg. What is the weight of the lightest package?

8. An object moving in a straight line travels 3m in the first second, 8m in the second second and 13m in the third second and so on. How far does it travel in 6 seconds? How far does it move in the 12th second?

---------------------------------------------------------------------------------------

| Solutions: Exercise 29.10 | |
|---|---|
| 1. 15 | 5. 0.5, 1 and 1.5 million |
| 2. $89000 | 6. $35000 |
| 3. $260000; $1520000 | 7. 0.25kg |
| 4. 10; $11125 | 8. 93m; 58m |

## GEOMETRIC SERIES OR PROGRESSION

Unlike an arithmetic sequence, a geometric sequence has a common ratio. Let us call the first term of an geometric sequence to be 'a' and the common ratio to be 'r'.

The sum $(S_n)$ of a geometric sequence with n terms and the last term 'l' can be written as:

$$S_n = a + ar + ar^2 + ar^3 + \ldots\ldots + ar^{n-2} + ar^{n-1} \text{ -----------(1)}$$

The $n^{th}$ term (or $T_n$) is $ar^{n-1}$ because all terms have 'a' but the power of 'r' of any term is one less than the number of the term (ie the third term has $r^2$, fourth term has $r^3$ and so on). Therefore, $T_n = ar^{n-1}$.

Multiplying each term in equation (1) by r and writing the result under the next term:

$$rS_n = ar + ar^2 + ar^3 + \ldots\ldots\ldots + ar^{n-1} + ar^n \text{ --------(2)}$$

(1) – (2) gives $S_n - rS_n = a - ar^n$

That is, $(1-r)S_n = a(1-r^n)$

Therefore, $S_n = \dfrac{a(1-r^n)}{1-r}$   (Valid only if r < 1)

When r is less than 1, the sum of a geometric series converges to a definite value when n is very large (ie when n tends to infinity). If r is less than 1 and n tends to infinity, $r^n$ will tend to zero. So the above formula when n tends to infinity becomes:

$$S_{infinity} = \dfrac{a}{1-r}$$

The formula for the sum of n terms of a geometric progression when r is greater than 1 is written as $S_n = \dfrac{a(r^n - 1)}{1-r}$  (Valid only if r > 1).

A geometric series is said to be divergent when r is greater than 1 and convergent when r is less than 1.

## GEOMETRIC MEAN(S)

Geometric mean of any two numbers is a number between the two numbers which along with the given number will form a **geometric progression**.

**Example:**

- Find the geometric mean of 2 and 32

Let the geometric mean be x. Then, 2, x and 32 will form a GP. That is, the common ratio between the first two and the last two numbers are equal.

Therefore, $\dfrac{x}{2} = \dfrac{32}{x}$

$x^2 = 64$   (by cross multiplication)

$x = \sqrt{64} = 8$

Therefore, the geometric mean is 8.

| The geometric mean of two numbers A and B = $\sqrt{AB}$ |
|---|

We can also insert more than one geometric mean between any two numbers so that all the numbers together form a GP.

---

**Example:**

- Insert three geometric means between 2 and 162

Let the three geometric means be $x_1$, $x_2$ and $x_3$. Therefore 2, $x_1$, $x_2$, $x_3$ and 162 form a GP. That is, a = 2 and $T_5$ = 162

$T_5 = ar^4$
$162 = 2 \times r^4$
$r^4 = 81$
$r = 3$

Therefore, $x_1$ = 6, $x_2$ = 18 and $x_3$ = 54

The three Geometric Means are 6, 18 and 54

---

**EXERCISE 29.11**--------------------------------------------------------------------------------

Find the eighth term and the sum of the first eight terms for each of the following Geometric Progressions:

1. 1, 2, 4, 8..............

2. 3, 6, 12, 24.............

3. 2, 6, 18, 54..............

4. 9, 3, 1, 1/3, ..............

5. 2, 1, ½. ¼.............

6. 5, −10, 20, −40 ................

7. 1/6, 1/8, 3/32, 9/128.............

8. 0.2, 0.8, 3.2, 12.8 ..............

--------------------------------------------------------------------------------

| Solutions: Exercise 29.11 | | |
|---|---|---|
| **1.** 128, 255 | **5.** | $\dfrac{1}{64}$, $3\dfrac{63}{64}$ |
| **2.** 384, 765 | **6.** | − 640, − 425 |
| **3.** 4374, 6560 | **7.** | $\dfrac{729}{32768}$, $\dfrac{58975}{98304}$ |
| **4.** $\dfrac{1}{243}$, $13\dfrac{121}{243}$ | **8.** | 3276.8, 4369 |

**EXERCISE 29.12**-----------------------------------------------------------------------

Using the formula $T_n = ar^{n-1}$, find the number of terms in each of the following Geometric Series:

1. $1 + 3 + 9 + \ldots\ldots\ldots + 729$

5. $1\frac{1}{2} + 4\frac{1}{2} + 13\frac{1}{2} + \ldots\ldots\ldots + 3280.5$

2. $2 + 4 + 8 + \ldots\ldots\ldots + 2048$

6. $2 + 0.6 + 0.18 + \ldots\ldots\ldots + 0.00486$

3. $8 + 4 + 2 + \ldots\ldots\ldots\ldots + 0.125$

7. $2 + 10 + 50 \ldots\ldots\ldots + 31250$

4. $\frac{1}{2}, \frac{1}{4}, 1/8 + \ldots\ldots\ldots\ldots + 1/256$

8. $1/8 - 1/4 + \frac{1}{2} + \ldots\ldots\ldots\ldots - 64$

| Solutions: Exercise 29.12 | | | |
|---|---|---|---|
| **1.** 7 | **3.** 7 | **5.** 8 | **7.** 7 |
| **2.** 11 | **4.** 8 | **6.** 6 | **8.** 10 |

**EXERCISE 29.13**-----------------------------------------------------------------------

Find the sum of the following Geometric Series;

1. $1 + 2 + 4 + \ldots\ldots\ldots$ to 9 terms

5. $1/2 + 1/6 + 1/18 + \ldots\ldots\ldots$ to infinity

2. $16 + 8 + 4 + \ldots\ldots\ldots$ to 8 terms

6. $1.5 + 0.3 + 0.06 + \ldots\ldots$ to infinity

3. $\frac{1}{4} + \frac{1}{2} + 1 + \ldots\ldots\ldots$ to 10 terms

7. $3 + 4 + 5\frac{1}{3} + \ldots\ldots\ldots$ to 7 terms

4. $3 - 6 + 12 - \ldots\ldots\ldots$ to 6 terms

8. $1.2 + 1.8 + 2.7 + \ldots\ldots\ldots$ to 6 terms

| Solutions: Exercise 29.13 | |
|---|---|
| **1.** 511 | **5.** $\frac{3}{4}$ |
| **2.** $31\frac{7}{8}$ | **6.** 1.875 |
| **3.** $255\frac{3}{4}$ | **7.** $58\frac{103}{243}$ |
| **4.** −63 | **8.** 24.94 |

163

**EXERCISE 29.14**----------------------------------------------------------------------------------------------

Insert the indicated number of Geometric Means between each of the following pairs of numbers:

1. 4 and 16 (1)    3. 3 and –96 (4)    5. 6 and 6144 (4)    7. 7 and 112 (3)

2. 2 and 162 (3)    4. 5 and $\dfrac{5}{64}$ (5)    6. $\dfrac{1}{2}$ and $\dfrac{1}{162}$ (3)    8. –4 and $\dfrac{1}{32}$ (6)

--------------------------------------------------------------------------------------------------------

| Solutions: Exercise 29.14 | |
|---|---|
| **1.** 8 | **5.** 24, 96, 384, 1536 |
| **2.** 6, 18, 54 | **6.** $\dfrac{1}{6}, \dfrac{1}{18}, \dfrac{1}{54}$ |
| **3.** –6, 12, –24, 48 | **7.** 14, 28, 56 |
| **4.** $\dfrac{5}{2}, \dfrac{5}{4}, \dfrac{5}{8}, \dfrac{5}{16}, \dfrac{5}{32}$ | **8.** 2, –1, $\dfrac{1}{2}, -\dfrac{1}{4}, \dfrac{1}{8}, -\dfrac{1}{16}$ |

**EXERCISE 29.15**------------------------------------------------------------------------------

1. The 5$^{th}$ term of a GP is $\dfrac{64}{81}$ and the 7$^{th}$ term is $\dfrac{256}{729}$. Find the first term and the common ratio.

2. The first term of a geometric sequence is 8 and the sum to infinity is 12. Find the common ratio.

3. The first term of a geometric sequence is 6 and the fourth term is –162, find the common ratio.

4. Find the 11$^{th}$ term of a GP whose 3$^{rd}$ term is 16 and the 7$^{th}$ term is 1.

5. Find the first term of a geometric progression whose third term is $2\dfrac{1}{4}$ and sixth term is $\dfrac{-16}{81}$

6. The sum of n terms of a GP is $2^n - 1$. Find the first four terms of the GP.

7. Find the 25th term of a GP if the first term is 0.1 and the common ratio is 2.

8. Find the 7$^{th}$ term and sum of the seven terms of the GP 0.005, 0.05, 0.5....

--------------------------------------------------------------------------------------------------------

## Solutions: Exercise 29.15

| | |
|---|---|
| **1.** $a = 4$; $r = \pm\dfrac{2}{3}$ | **5.** $\dfrac{729}{64}$ |
| **2.** $r = \dfrac{1}{3}$ | **6.** 1, 2, 4, 8 |
| **3.** $r = -3$ | **7** 1677721.6 |
| **4.** $\dfrac{1}{16}$ | **8.** 5000; 5555.555 |

## EXERCISE 29.16------------------------------------------------------------------

1. A land purchased for $120000 appreciates in value at the rate of 6% per annum. What would be its value after 8 years?

2. A ball is dropped from a height of 64 metres. When it rebounds it reaches to a quarter of the previous height each time. Find the total distance it travels.

3. The population of a town in 2010 was 2100000. If it increases by 10% each year, what would be the population in 5 years?

4. A Motor car bought for $45000 depreciates in value at the rate of 20% per year on the declining value. What would be its value at the end of 6 years?

5. Assuming that the rate of inflation is steady at 4% per year compounded annually, how much will a car costing $32000 now cost in 5 years?

6. A company's labour force is increasing at an annual rate of 1.7% per year. If the company's labour force is 78 at present how many will it be expected to employ in four years?

7. The bacterial count in a certain culture increasing from P to A at an average increase per day of r% is given by the equation $A = P(1 + r)^n$ where n is the total number of days. What will be the value of r if P = 2000, A = 8000 and n = 3. What will be the bacterial count on the second day?

8. A water tank with a capacity of 120 litres was filled and there was a leak which drained the water at the rate of 5% of the water remaining every minute. How long will it be before the tank is half full?

------------------------------------------------------------------

## Solutions: Exercise 29.16

| | | | |
|---|---|---|---|
| **1.** | $180435.63 | **5.** | $37435.47 |
| **2.** | 85.33m | **6.** | 82 |
| **3.** | 3074610 | **7.** | r = 0,587; 3174 |
| **4.** | $14,745.60 | **8.** | 13.51 minutes |

There are several related series that are known as the **binomial series**.

> The most general is: $(a + b)^n = {}^nc_0.a^n + {}^nc_1.a^{n-1}b + {}^nc_2.a^{n-2}b^2 \ldots {}^nc_{n-1}.a.b^{n-1} + {}^nc_n.b^n$

The $r^{th}$ term of this series can be written as: ${}^nc_{r-1}.a^{n-r+1}b^{r-1}$. The series is true for any positive integral value of n. Note that the coefficients of the terms are ${}^nc_0, {}^nc_1, {}^nc_2$ and so on.

The coefficients can also be obtained using the Pascal's Triangle. The ${}^nc_r$ method saves the trouble of writing out all the lines of Pascal's Triangle to find the coefficients.

The binomial series is true for negative values and fractional values of n as well if $|b/a| < 1$.

---

**Examples:**

- Expand $(1 + x)^5$

$(1 + x)^5 = {}^5c_0(1)^5 + {}^5c_1(1)^{5-1}x^1 + {}^5c_2(1)^{5-2}x^2 + {}^5c_3(1)^{5-3}x^3 + {}^5c_4(1)^{5-4}x^4 + {}^5c_5x^5$

$(1 + x)^5 = 1 + 5x + \dfrac{5 \times 4}{2}x^2 + \dfrac{5 \times 4 \times 3}{3 \times 2 \times 1}x^3 + 5x^4 + x^5$

$(1 + x)^5 = 1 + 5x + 10x^2 + 10x^3 + 5x^4 + x^5$

- Expand $(2 - a)^6$

$(2 - a)^6 = {}^6c_0(2)^6 + {}^6c_1(2)^5(-a)^1 + {}^6c_2(2)^4(-a)^2 + {}^6c_3(2)^3(-a)^3 + {}^6c_4(2)^2(-a)^4 + {}^6c_5(2)^1(-a)^5 + {}^6c_6(2)^0(-a)^6$

$(2 - a)^6 = 64 - 192a + 240a^2 - 180a^3 + 60a^4 - 12a^5 + a^6$

- Write down the $4^{th}$ term of $(p + q)^7$

The $4^{th}$ term $= 7c_3p^4q^3$

$= \dfrac{7 \times 6 \times 5}{3 \times 2 \times 1}p^4q^3$

$= 35p^4q^3$

- Find the value of $\sqrt{1.06}$

$\sqrt{1.06} = (1 + 0.06)^{1/2}$

$= {}^{1/2}c_0 (1)^{1/2} + {}^{1/2}c_1 (1)^{-1/2}(0.06) + {}^{1/2}c_2 (1)^{-3/2}(0.06)^2 + {}^{1/2}c_3 (1)^{-5/2}(0.06)^3 + \ldots\ldots$

$= 1 + \dfrac{1}{2}(0.06) + \dfrac{0.5 \times -0.5}{2 \times 1}(0.0036) + \dfrac{0.5 \times -0.5 \times -1.5}{3 \times 2 \times 1}(0.000216) + \ldots\ldots$

$= 1 + 0.03 - 0.00045 + 0.000013 + \ldots\ldots$

$= 1.02956$

## EXERCISE 29.17 -----------------------------------------------------------------

Write out the expansion of each of the following using the ${}^nc_r$ method to find the coefficient of each term:

1. $(1 + x)^3$     3. $(m - n)^5$     5. $(2a - 3b)^7$     7. $(2 + 0.05)^3$

2. $(2 + x)^4$     4. $(a + 2)^6$     6. $(1 + 0.2)^4$     8. $(3p + 1)^6$

-----------------------------------------------------------------

| Solutions: Exercise 29.17 | |
|---|---|
| **1.** $1 + 3x + 3x^2 + x^3$ | **5.** $128a^7 - 1344a^6b + 6048a^5b^2 - 15120a^4b^3$ $+22680a^3b^4 - 20412a^2b^5 + 10206ab^6 -$ $2187b^7$ |
| **2.** $16 + 32x + 24x^2 + 8x^3 + x^4$ | **6.** $1 + 0.8 + 0.24 + 0.032 + 0.0016$ $(= 2.0736)$ |
| **3.** $m^5 - 5m^4n + 10m^3n^2 - 10m^2n^3 + 5mn^4 - n^5$ | **7** $8 + 0.6 + 0.015 + 0.000125$ $(= 8.615125)$ |
| **4.** $a^6 + 12a^5 + 60a^4 + 160a^3 + 240a^4 + 192a^5$ $+ 64$ | **8.** $729p^6 + 1458p^5 + 1215p^4 + 540p^3 +$ $135p^2 + 18p + 1$ |

**EXERCISE 29.18**--------------------------------------------------------------------------------

1. Write down the 8th term in the expansion of $(2x + 3y)^7$

2. Using the binomial expansion find the value of $\sqrt{1.02}$ correct to 5 decimal places.

3. Expand $(1 + 0.08)^6$ giving the answer corrected to two decimal places.

4. What is the $5^{th}$ term in the expansion of $(1 + 0.04)^{-3}$

5. The probability that in a family of four children there will be three boys is given by $P_r(3 \text{ boys}) = {}^4c_3(\frac{1}{2})^1(\frac{1}{2})^3$. Calculate this probability.

6. A machine produces 20% defective product. The probability that out of 4 items chosen at random 2 are defective is given by ${}^4c_2 (0.2)^2(0.8)^2$. What is this probability?

7. Find the probability that in five tosses of a fair die 3 appears twice. [Hint: ${}^5c_2(5/6)^3(1/6)^2$]

8. Using the binomial series find the $4^{th}$ term in the expansion of $(1 + x^2)^{1/2}$

--------------------------------------------------------------------------------

| Solutions: Exercise 29.18 | |
|---|---|
| **1.** $2187y^7$ | **5.** $\frac{1}{4}$ |
| **2.** 1.00995 | **6.** 0.1536 |
| **3.** $1 + 0.48 + 0.096 + 0.01024 +$ $0.0006144 + 0.0000196608 +$ $0.000000262144 (= 1.59)$ | **7.** $\frac{625}{3888}$ |
| **4.** 0.0000384 | **8.** $\frac{1}{16}x^6$ |

# INDEX

Absolute value, 96
Addition, 7,12
Algebraic expression, 3
Antilogarithm, 52
Arithmetic mean, 156
Arithmetic sequence, 147
Binomial series, 166
BOMDAS rule, 34
Brackets, 37
Characteristic, 51
Coefficient, 2
Common factors, 59
Common logarithm, 51
Complex variables, 3
Conjugate, 45
Constant, 1
Converting logarithms, 52
Cube root, 39
Cubic functions, 132
Discriminant, 121
Dividend, 25
Divisor, 25
Elimination method, 114
Exponent, 48
Exponential functions, 98
Factor theorem, 128
Factorial notation, 151
Factorization, 59
Fibonacci sequence, 145,147
Finite sequence, 148
Fractional surd, 45
Fractions, 76
Geometric mean, 161
Geometric progression, 161
Geometric sequence, 147
Growth patter, 144
Highest common factor, 70
Indices, 47
Index, 48
Inequality signs, 103
Inequation, 103
Infinite sequence, 148
Intersection point of functions, 132
Interval notation, 105
Like terms, 4
Long division, 27
Logarithm, 51

Lowest common multiple, 73
Mantissa, 51
Maximum value of the function, 131
Minimum value of the function, 131
Natural logarithm, 52
$^{N}c_{r}$ notation, 151
Number line, 104
Number patterns, 144
Numerical sequence, 145
Ordinals, 34
Pascal's Triangle, 19
Partial fractions, 137
Pattern, 144
Polynomial, 126
Polynomial functions, 130
Product, 14
Product variables, 2
Quadratic expression, 63
Quadratic equation, 119
Quadratic formula, 121
Quadratic function, 131
Quartic functions, 132
Quotient, 25
Rationalization, 45
Remainder, 27
Remainder theorem, 128
Repeating pattern, 144
Recursive pattern, 145
Rules of indices, 48
Roots, 119
Rule, 145
Sequences, 147
Series, 153
Sigma notation, 153
Simple equation, 84
Simple linear equation, 84
Simultaneous equations, 111
Square root, 39
Substitution, 56
Substitution method, 111
Subtraction, 10,12
Surd, 43
Term, 3
Unlike terms, 4
Variable, 1
Whole number, 43

Printed in the United States
By Bookmasters